D0539744

100
IDEAS
THAT CHANGED
THE WORLD

100
IDEAS
THAT CHANGED
THE WORLD

Our most important discoveries,
selected by our greatest minds

JHENI OSMAN

1 3 5 7 9 10 8 6 4 2

Published in 2011 by BBC Books, an imprint of Ebury Publishing
A Random House Group company

Copyright © Woodlands Books Ltd 2011

All rights reserved. No part of this publication may be reproduced, stored in a retrieval system,
or transmitted in any form or by any means, electronic, mechanical, photocopying, recording
or otherwise, without the prior permission of the copyright owner.

The Random House Group Limited Reg. 954009

Addresses for companies within the Random House Group can be found at
www.randomhouse.co.uk

A CIP catalogue record for this book is available from the British Library

ISBN 978 1 84 9901963

The Random House Group Limited supports The Forest Stewardship Council (FSC®), the
leading international forest certification organisation. Our books carrying the FSC label are
printed on FSC® certified paper. FSC is the only forest certification scheme endorsed by
the leading environmental organisations, including Greenpeace. Our paper procurement
policy can be found at www.randomhouse.co.uk/environment

Commissioning editor: Albert DePetrillo
Project editor: Joe Cottington
Copy editor: Eddie Mizzi
Designer: O'Leary & Cooper
Illustrations: Richard Palmer
Production: Phil Spencer

Printed and bound in the UK by CPI Mackays, Chatham, ME5 8TD

To buy books by your favourite authors and register for offers, visit www.randomhouse.co.uk

CONTENTS

The Natural World

The Universe **87**

The World of Physics 123

The Chemical World 175

The Living World   **203**

The Medical World 249

The World of Engineering   327

The World of Technology   361

INTRODUCTION

On 25 May 1961, JFK announced his bold vision to send astronauts to the Moon before the end of the decade. Eight years later, as the *Apollo 11* lunar module touched down, Neil Armstrong uttered the words: "Houston, Tranquility Base here. The Eagle has landed." That iconic moment was the culmination of thousands of years of technological progress.

Discoveries rarely come about by someone leaping out of the bath shouting "Eureka!" More often they are a melding of minds and ideas that are in the scientific consciousness at that time, and are then enabled by technological advancements such as the printing press. As Isaac Newton said: "If I have seen further it is by standing on the shoulders of giants."

Often it comes down to the right time and the right place. Sometimes it's sheer chance, for example when Alexander Fleming discovered penicillin (see page 260). Occasionally, it's about sticking to your guns with an unpopular idea or thinking the seemingly unthinkable, as Copernicus did when he proposed that the Earth orbits the Sun (see page 95). Indeed, one day our descendants may laugh at us for believing a certain theory or taking for granted a particular physical law.

While writing this book, I was often asked what I thought was the biggest idea that changed the world. There are so many incredible discoveries and brilliant inventions to pick from: how do you decide?

WHAT I FOUND OUT WHILE WRITING THIS BOOK

- Urine is recycled into drinking water on the International Space Station, which saves transporting water up to the ISS at a cost of $40,000 per gallon. See page 335.
- If you took all the DNA in one human cell and laid the chromosomes out end to end, they would reach nearly two metres. See page 241.
- The well-known tale that Isaac Newton discovered the concept of gravity by watching an apple fall from a tree is almost definitely untrue. See page 144.
- About 90 million mobile phones are lying unused in homes throughout the UK, amounting to 11,250 tonnes – five times the weight of the London Eye. See page 378.
- The megathrust earthquake that caused the Boxing Day tsunami in 2004 displaced 30 km³ of water that created a series of waves so powerful that they travelled round the world for 40 hours. See page 57.
- The Hubble space telescope is the size of a large school bus, and travels at 8 km per second, taking just over an hour and a half to complete one orbit of the Earth. See page 92.
- If tin foil gets stuck in your teeth, a tiny electric current is generated between the foil and any metal fillings you have. See page 153.

As a young girl I had two posters on my wall. One was an illustrated poem, from the Natural History Museum, of the complete history from the dinosaurs to the 1980s (or as comprehensive as an A3 poster allows). The other was a picture of the Space Shuttle. To mini-me, the idea that there were astronauts rocketing into space inside that craft was mind-boggling. And so, for that reason, I would choose space travel as the big idea and the Moon landing as the most iconic moment in our history.

The outstanding people I interviewed while writing this book have each picked out one invention, theory or discovery that they feel truly changed the world. While the book is written by me, when one of the experts contributed their thoughts on why the idea was so revolutionary, their words appear as a quote.

The experts sometimes surprised me with the ideas that they chose, but the more I researched their choices, the more I would understand their decision and how the idea made its impact on society. On the opposite page are listed just a handful of things I found out while writing this book. I hope you enjoy reading it as much as I enjoyed writing it.

Science
in Society

THE SCIENTIFIC METHOD

Seeking the truth

Nominated by **Professor Brian Cox**, professor of physics at Manchester University, Royal Society University Research Fellow and BBC presenter

According to legend, one day in 1612 the Italian scientist Galileo clambered up to the top of the Leaning Tower of Pisa, dropped two weights from the top and showed that despite having very different masses, they hit the ground at the same moment (see page 144). It was a startling, counter-intuitive discovery and completely refuted the teachings of Aristotle, which then held sway in the academic world.

While the Leaning Tower story itself may be the result of several hundred years of embellishment and exaggeration, there's no doubt that Galileo performed experiments aimed at debunking Aristotle's views of how gravity works. That makes Galileo a pioneer in what has become known as the 'Scientific Method': the formulation of a hypothesis, followed by an experiment to test the hypothesis and the drawing of conclusions. It was an approach to understanding the world which Aristotle had eschewed – simply because he did not believe that the imperfect world of reality could have any bearing on the perfect world of abstract ideas. For him, certain knowledge could only be obtained by deduction from obvious facts, and experimentation was irrelevant.

Today, following Galileo, we think differently. The laws of nature, and the scientific theories that explain them, arise from generalizations that are inferred from observations in the real world – a procedure termed 'induction'. For a theory to be deemed scientific, what it asserts must be supported by evidence that can be physically observed and measured. Scientific theories are carefully constructed models of the world that attempt both to explain the observed phenomena and make predictions that can be tested by experiment. It is these so-called 'crucial experiments' that determine the fate of a scientific theory.

If a theory passes an experimental test, it is in a sense corroborated, but it can never be confirmed or proved ultimately 'true', for no amount of observation or experiment can establish the truth of a theory with absolute certainty. There is always the possibility that one day a crack will be found in the edifice of the theory. If the theory fails the test, however, it faces a much harder challenge. Sometimes the experiment can be shown to be faulty, either in execution or design. Even so, a single, well-designed experiment can be enough to dismiss a theory outright. As Einstein supposedly once put it: "No amount of experimentation can ever prove me right; a single experiment can prove me wrong." This notion led the Viennese philosopher of science Karl Popper to the concept of 'falsifiability' as the defining characteristic of true science: a theory is scientific only if it can – at least in principle – make predictions that can be shown to be false. And the best theories are those that withstand continued attempts to refute them by rigorous testing.

ALHAZEN AND THE SCIENTIFIC METHOD

Abu Ali al-Hasan ibn al-Haytham (aka Alhazen) was born around AD 965 in Basra, in what is now Iraq. He was a true polymath, experimenting and forming theories in everything from physics to astronomy to psychology. His groundbreaking book on optics was the definitive work on the subject for 650 years, until Newton arrived on the scene. Despite a stint under house arrest after displeasing the Caliph by promising more than he could deliver on an engineering project to dam the Nile, he wrote over 200 manuscripts during his lifetime.

But Alhazen is probably best known for being a pioneer of the Scientific Method – his rigorous system of prediction and observation laying the foundations for scientific experimentation. For that alone he deserves the accolades of having an asteroid and crater on the Moon named after him, as well as his status as an Iraqi national treasure: his face sits on the 10,000 dinar banknote.

Galileo's appeal to observation and experiment to adjudicate on a scientific theory formed the basis of a new approach to studying nature. His revolution consisted in establishing induction as the logical method of science. This is not to say that scientists before him had not considered experimentation as being important to science, for many had. Indeed, the great Arab scholar Alhazen is considered by some to be the father of the scientific method for the surprisingly modern way in which he conducted his research on optics.

Finally, the scientific method relies on open and free communication among scientists. Secrecy has no place in science. In the words of the Austrian-born British biochemist Max Perutz: "True science thrives best in glass houses, where everyone can look in. When the windows are blacked out, as in war, the weeds take over; when secrecy muffles criticism, charlatans and cranks flourish."

THE PATENT

Without a patent, invention is futile if you want to cash in

Nominated by Sir James Dyson,
inventor and entrepreneur

"While there are undoubtedly issues with the patent system, a patent itself is the cleverest of inventions: it keeps the idea safe while laying it open to the public," says Dyson.

A patent is something new, applicable to industry and not so obvious that anyone could have thought of it. Although patent law varies from country to country, it's generally the case that if you gain a patent it gives you the right to exclude others from making, using, selling or importing your invention – for the term of the patent, which is usually 20 years. But, crucially, a patent allows you to sell or licence your invention should you wish to do so.

"Ideas are an inventor's lifeblood. But without a patent, an idea is unprotected – and worth nothing. Research and development of new technology thrives on secure intellectual property."
James Dyson

History of the Patent

The idea of a patent goes back as far as 500 BC, when people in the Greek city of Sybaris, in modern-day southern Italy, were encouraged to discover new ways of making luxurious goods. In return they'd reap the profits from the invention for a year.

Through the ages, all sorts of patents were applied for and obtained. The idea of a patent that prevented others from using it was developed in 1474 in the Republic of Venice. The concept of a patent having to be a new invention was decreed by King James I in 1623. Then in the reign of Queen Anne (1702–14), lawyers in the English Court decreed that a written description of the invention had to be formally submitted.

Bitter disputes and court cases have raged throughout more recent history between inventors trying to secure intellectual property rights. Take the case of the telephone.

Antonio Meucci is now credited with inventing a proto-telephone. The Italian actually obtained a patent for his device in 1871, but, in 1874, lack of cash meant he couldn't afford the $250 needed to secure the patent. Alexander Bell shared a lab with Meucci and spotted his chance. He improved Meucci's device and found the money to patent it in 1876.

So Meucci was stung by not having enough cash to keep his patent. But he's not the only one. Even today's highly successful inventors have missed out.

Cashing in

Dyson is famous for inventing the bagless vacuum cleaner and became rich from it – he's reportedly worth over £1.1 billion. But it's not always been plain sailing for him: "When I started out, I once lost a UK patent for a valve on my vacuum cleaner because I couldn't afford it. Soon it was being used by one of my competitors."

It's a Catch-22, you've got to have the money to patent an invention, but so often amateur inventors don't have the cash to secure their idea. As Dyson says: "For start-ups and small businesses, getting a patent can be painstaking and expensive. But worth fighting for."

TIMELINE OF PATENTS OBTAINED

1849	The safety pin	Walter Hunt
1869	The roller skate	Isaac Hodgson
1876	The telephone	Alexander Graham Bell
1880	The electric lightbulb	Thomas Alva Edison
1886	Dish-washing machine	Josephine Cochrane
1888	Camera with roll film	George Eastman
1895	Alternating current electricity	Charles Steinmetz
1906	Air conditioning	Willis H. Carrier
1911	Flying machine	Alexander Graham Bell and G. H.Curtis
1941	Toothbrush	Frank E. Wolcott
1944	Lipstick	Eleanor Kairalla
1955	Velcro	George de Maestral
1966	The Batmobile design	George Barris
1968	Weak adhesive used for Post-it notes	Spencer Silver
2006	A device for the treatment of hiccups	Philip Charles Ehlinger, Jr
2006	USB memory stick	Chih-Chien Lin

HUMAN LANGUAGE

The evolution of our ability to speak

Nominated by **Dr Penny Fidler**, neuroscientist and CEO
of the Association for Science and Discovery Centres

*"The development of the human
brain to create and process
complex spoken language has been
fundamental to the advancement of
our species. This relative revolution,
in evolutionary terms, has allowed
us to share intricate ideas with one
another, to pass on knowledge and
learning, to discuss the past and
future and share plans with whole
populations, and, of course, to
express how we think and feel."*
Penny Fidler

Obviously, no fossils exist to
show how we evolved to use
language. To work out when
it first developed from the
grunts of our ape ancestors,
scientists have looked at
indirect evidence, such as
comparing the behaviour of
animals and humans.

The difference between
an animal communicating
and a human language is
that emotions can actually
be explained, rather than
just signalled. For example,
a cat meowing might show

how distressed the animal is, but it's not able to explain to another cat the exact reason for this. Although some species can convey more meaning than others – bees dance in certain ways to indicate flower location, for example – it is not considered a language, because there are limits to what they can convey.

True language requires the ability to learn, understand, and then produce new words. Some animals can do some of these, but we're the only ones that can do all three. By comparing animal and human language traits, we can work out when human language might first have developed. The jury is still out on this, but research suggests that our ability to understand sounds developed in a mammalian ancestor before the ability to produce complex sounds.

Creating complex sounds was a massive breakthrough in the development of language. Our closest common ancestor – the chimpanzee, from which we diverged about 6–7 million years ago – isn't able to do this. But what physiological trait allowed us to develop speech?

Babies can breathe through their nose while suckling. By around three years old, their larynx has descended down their throat. A lower larynx was thought to be the reason that humans could produce complex sounds in speech, but a number of other animals, from tigers to koalas, have been found to have a descended larynx. So our unique ability to speak must be due to something else.

The human brain is incredibly complex. Studies on the brainstem have found that neurons responsible for controlling our larynx, tongue and lips are directly linked to the motor cortex, which is involved in planning, controlling and carrying out movements. Other primates lack these connections, so it seems that our ability to speak is due to our complex brains, while our ability to make coherent sentences seems to be down to our genes.

DID YOU KNOW?

- English has the largest number of distinct words, around 250,000, and the Cambodian language, Khmer, has the largest alphabet, with 74 different letters.

- Papua New Guinea is the country with the most living languages – over 800.

- More than 90 per cent of people in the world speak a language from the top ten most common language families.

A variant of the gene *FOXP2* has been shown to exist only in humans. The gene is vital for forming memories in the brain areas responsible for patterns of movement that are involved in complex speech. If this gene is disrupted in someone, they suffer from severe speech problems. DNA analysis has found that Neanderthals had this gene. Contrary to the traditional view that we have of them as grunting, incomprehensible ape-like creatures, it seems that, in fact, they may have been able to speak.

Some scientists believe that language existed before Neanderthals. *Homo erectus* had a much larger brain than its predecessors and made more advanced tools. But their technological development seems to have plateaued, suggesting that any proto-language they spoke wasn't sufficiently complex to enable further technological progress. Before *Homo erectus*, it's likely that gestures as opposed to speech originally governed language; but with increased tool use and the need to communicate in the dark, speech developed.

Much is still unknown about the evolution of language, and, also, how it will evolve in the future. Throughout history, words have fallen in and out of the dictionary, and different languages have dominated in different eras. One of the big questions currently challenging linguists is, will English continue to be the main second language for most cultures? Only time will tell.

WRITING

The invention that enabled science to flourish

Nominated by **Dr Yan Wong**, evolutionary biologist and presenter on BBC science show *Bang Goes the Theory*

'Pictographs' – also known as 'pictograms' – could be considered the first form of writing. But they are different from modern-day alphabets in that the picture conveys its meaning by looking like a physical object.

"Writing is why ideas aren't lost but built upon. So it's the reason we can do science at all. In fact, writing has caused one of the 'major transitions' in evolutionary biology. With language and writing, instead of genes being the sole container of evolvable information, ideas can also be transmitted, changed, and so evolve."
Yan Wong

Cave paintings, using early versions of pictographs, have been discovered in France, dating from around 32,000 years ago.

But it wasn't until around 8000 BC that pictographs started to be pressed into clay tablets. The pictographs were now transportable, but the characters were still somewhat randomly aligned.

Around 3000 BC, the cuneiform script was developed by the Sumerians, who lived in what is now Iraq. This script was a series of pictographs inscribed on clay tablets using a stylus – a sharpened tip of a reed. Initially, the pictographs were laid out in vertical columns, but later they were read from left to right.

Portable Papyrus

As people felt the need to spread and share information, they devised lighter, more portable objects to write on. The ancient Egyptians pressed papyrus plants to make scrolls, on which they wrote in ink made from a mixture of soot, gum and water. The Romans wrote on stretched animal skin, known as parchment or vellum. Then, around the second century AD, the Chinese invented paper – and a clever way of printing onto it.

By carving whole texts on wooden blocks, then brushing the blocks with ink, the Chinese, Koreans and Japanese were able to press reams of paper onto the inky blocks to reproduce texts again and again. Different pages of the text could then be sewn together into books.

Even after woodblock printing was invented, the written word remained the domain of the elite – the rich and the clergy. The sheer cost of books rendered most of the population illiterate. But around 1440, Johannes Gutenberg revealed a major breakthrough that he'd been working on – movable type. This revelation led to the invention of the printing press (see page 39).

The Future of Writing

The future of writing and language remains uncertain. The digital age of typing, texting and emoticons have changed how we communicate and the language that we use to do so. Short-hand expressions such as 'LOL' (which is now in the Oxford English Dictionary) and the emoticon ;) are even starting to creep into the business world in texts and emails to work colleagues.

Recent research by scientists at the University of Stavanger in Norway and University of Marseille in France shows that writing something down makes us remember it better. But in this information age, have smartphones and Google replaced our need to remember certain things? In the future, could our brains, gadgets and search engines be so integrated that memory and writing become a thing of the past?

"It's unlikely that writing will completely disappear," says Wong. "Anyway, it won't be the end of the information contained in the writing. In fact, thanks to modern technology, the replication of information tends to go from strength to strength. So the evolution of ideas continues unceased, and is even revolutionised by modern technology."

PAPER

From pounding rags to pulping wood

Nominated by **Dr Nikolaos Psychogios**, Harvard Medical School

As the written word materialised, people searched for lightweight portable objects to record ideas on, rather than cumbersome clay tablets. For centuries, the ancient Egyptians had pressed together layers of pith from the papyrus plant to create a material to write on, but it was expensive to produce. With the growing demand to communicate longer texts more frequently across larger distances, a cheaper alternative was needed.

In 2006, at Fangmatan in the northeast Gansu Province of China, specimens of paper were

> *"The invention of paper enabled people to store thoughts and record their words in something small and portable that helps exchange information and ideas. It revolutionised the way we teach, learn and communicate."*
>
> Nikolaos Psychogios

discovered bearing Chinese characters. They had been used by the military and were dated to the first century AD. In the second century, court official Cai Lun came up with a more modern method of paper-making, combining plant fibres with hemp waste and pieces of fishing net and old rags. Rags might seem a slightly odd thing to add to the mix, but it is thought someone may have come up with the idea after watching rags being washed and pounded before the matted fibres were laid out on a mat.

The Chinese were secretive about their new invention. Legend has it that, after the Chinese army was defeated in AD 751 in a battle in Talas in modern-day Kyrgyzstan, two prisoners were tortured until they revealed how paper was made, and subsequently a paper mill was built in Samarkand in modern-day Uzbekistan.

Over the centuries, paper-making technology gradually made its way along the network of trade routes, known as the 'Silk Road'. In Baghdad, the elaborate paper-making process was refined so that paper could be mass-produced, and the traditional Chinese method of using a pestle and mortar to pound the pulp by hand was replaced with a trip hammer.

Paper-making technology didn't make it to Europe for many centuries. It was adopted in Spain and Sicily in the tenth century, but only reached northern Europe at the turn of the fifteenth century. The first known paper mill in England was built near Stevenage in 1490.

Paper continued to be pricey, though, and it wasn't until the nineteenth century that two ingenious ideas drove down the cost.

From Rags to Riches

The clever idea of a continuous paper-making machine that created rolls instead of sheets of paper was first patented in 1799 by the Frenchman Nicholas Robert. Controversy followed after his boss Léger Didot claimed he himself had invented it and contacted stationer Henry Fourdrinier about financing its construction.

Fourdrinier and his brother built two of these machines. They sold one of these so-called 'Fourdrinier machines' to Tsar Nicholas II of Russia, in return for £700 every year for ten years. The tsar didn't stick to his side of the bargain, though, and only paid up after years of chasing, by which time the brothers had gone bankrupt as they struggled to protect the patent.

Another key breakthrough in paper-making was the idea of using wood pulp instead of old rags. Matthias Koops had investigated the idea in the early 1800s, but, despite receiving financial support from the royal family for building his machine, went bankrupt. Recognising its potential, two inventors, the Canadian Charles Fenerty and the German Friedrich Gottlob Keller, separately experimented with wood pulp during the 1830s and 1840s and announced their findings in 1844. Fenerty showed a sample of his paper to Halifax's leading newspaper, while Keller tried to stoke the interest of the German government. Neither man made much money from his invention, though: Fenerty never patented his machine, while Keller sold his to Heinrich Voelter for a measly £80.

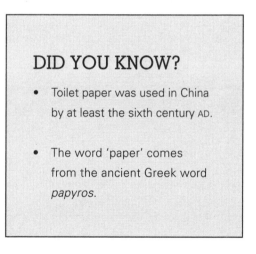

DID YOU KNOW?

- Toilet paper was used in China by at least the sixth century AD.

- The word 'paper' comes from the ancient Greek word *papyros.*

By the end of the nineteenth century, wood had replaced rags completely, and many mills were making a fortune from paper-making. From rags to riches, the history of paper spans many centuries and many continents – and what a key invention it has been. "Without paper, knowledge and science could not be precisely imparted," says Psychogios. "And the order of society is maintained, because laws exist on paper."

THE PRINTING PRESS

Taking the written word to the masses

Nominated by **Dr Paul Parsons**, science writer and author of *Science in 100 Key Breakthroughs*

After 20 years travelling all over Asia, merchant Marco Polo returned home to Venice laden with wares to trade and ideas to share. In 1298, Venice was at war with Genoa and Polo got caught up in the middle of it. In the Battle of Curzola, he was captured and locked up for three months. While in prison, he recounted his travels through Asia to fellow cell mate Rustichello da Pisa.

When the Venetians were finally released, Rustichello wrote about his cell mate's adventures

"Not only did the printing press and the easy availability of literature encourage people to read and write, but it was also a rapid and efficient way to cast text into print and then mass-produce it. It brought about the birth of news media: people could now read the news on a daily basis and be informed. It was a body blow in the fight to replace ignorance with reason."

Paul Parsons

in *The Travels of Marco Polo*, recounting stories of Polo's reputed life in the court of the Mongol leader Kublai Khan and also the existence of things such as paper money. The book omitted a number of key details from Chinese culture – the practice of foot-binding, eating with chopsticks and drinking tea – none of which existed in Europe at the time. It also didn't mention the difference between the Chinese script and Roman alphabet, nor did it talk of the technique of woodblock printing, which the Chinese had been doing for centuries.

As far back as the third century AD, carved blocks of wood were used by the Chinese to print patterns on fabric. Pi Cheng is credited with inventing movable wooden type in 1045, where one face of a block of wood was carved by hand to leave a raised character. By the thirteenth century, people in Korea started placing a layer of metal on to the characters to make them more durable. The thousands of ideographs in the Chinese language meant the process of printing was complex. What was needed was a language with fewer characters.

Legend has it that, although printing was never mentioned in *The Travels of Marco Polo*, the merchant realised how much simpler woodblock printing would be if the Roman alphabet with its limited number of characters were used, and so, reportedly, he introduced the idea to Europe.

For the next couple of centuries, everything from books to playing cards were printed using the laborious woodblock method. Then, in 1440, the son of a rich German merchant revealed a groundbreaking invention that he'd been working on in secret for years.

Printing Revolution

Johannes Gutenberg was preparing for showcasing polished metal mirrors that he and colleagues had created for an exhibition in the

German city of Aachen in 1439. At the last minute, the exhibition was cancelled and the investors came knocking. Instead of stumping up the cash to get the investors off his back, he promised to share with them the secret invention that he'd been working on.

Reportedly, the invention Gutenberg revealed was his printing press – a wooden machine which combined the idea of block printing with the technology used in a screw press for making olive oil and wine. Paper, mounted on a flat board called a 'tympan', was pressed down onto another flat board, a 'forme', which held inked type securely in place. The individual letters of the type were made by pouring the elements lead, tin and antimony into copper moulds, so the letters were quicker to make and more durable than hand-carved wooden ones.

The device completely changed the face of printing. It could produce as much material in one day as a scribe could produce in one year. Rather arrogantly, Gutenberg said of his invention: "Like a new star, it shall scatter the darkness of ignorance." But he was right: by the end of the sixteenth century, 150 million books had been published across Europe.

The Bible was one of the first books Gutenberg produced, and initially it dominated the book market. Gradually other texts became available, including science books such as Copernicus's *De Revolutionibus* (see page 95). Where once the printed word had been the domain of the wealthy and the clergy, Gutenberg's printing press opened up a new world to the literate.

"For those living at the time, the invention of the printing press must have been like the invention of the internet during the late twentieth century," says Parsons. "Suddenly a vast body of information became accessible – albeit to those who were literate. But that didn't matter because, just like the internet has encouraged people to become savvy with computers, so the printing press and the easy availability of literature that it brought fostered education and encouraged people to read and write."

Speed Printing

A faster press meant cheaper books, so people were always on the lookout for ways to improve the machine. In 1811, German inventors Friedrich Koenig and Andreas Bauer demonstrated their high-speed press driven by steam, which could print more than 15 pages a minute. Three years later, on 29 November 1814, *The Times* was the first newspaper to be printed on it.

Half a century on, clockmaker Ottmar Mergenthaler was asked by venture capitalist James Clephane and colleague Charles Moore to find an even quicker way to print legal documents. Mergenthaler realised that operators didn't need to place each precast metal letter one by one, but instead metallic letter moulds could be assembled in a long line. An operator then pressed buttons specific to certain letters to select whichever ones were required. These letters were then cast as a single piece of metal, known as a 'slug'.

This Linotype machine – 'line-o'-type' – speeded up the printing process dramatically. Before 1886, no newspaper had more than eight pages. But, in July of that year, a Linotype machine was installed in the offices of the *New York Tribune* and the age of mass-produced long-length papers began. The *New York Tribune* is no longer in publication, but today *The Times* sells about 450,000 copies every day.

MONEY

From prostitution to dealing on the
stock exchange, trading has been
around for thousands of years

Nominated by **Professor Dan Ariely**, behavioural economist
and author of books including ***The Upside of Irrationality***

Putting a finger on exactly
when bartering began is hard.
Humans have been doing it for
thousands of years, enabling
specialised professions to
develop. So someone growing a
certain crop can exchange with
another who has specialised in
producing a particular type of
fruit. Bartering allowed people
to be not just a jack of all trades,
but also master of one.

In the animal kingdom,
bartering of a sort happens in

*"We so often deal with money that
we overlook its importance. But
think for a moment what your life
would be like without it. Without
money we would not be able to
trade as easily (bartering is very
limited), we could not save for
retirement or for a rainy day, we
could not specialise and we could
not take on any large projects.
And, of course, there would be no
universities and no academics."*
Dan Ariely

symbiotic relationships. For example, the Egyptian plover gets a free meal by cleaning the teeth of the Nile crocodile, and in return the croc maintains its dental hygiene.

Animals aren't known to barter an actual object for another object, though. Studies on chimpanzees showed they were happy to trade fruit for tokens with humans, but they wouldn't do this between chimps, as tokens don't exist in the natural world. In 2008, researchers from Georgia State University, the University of California, and the UT MD Anderson Cancer Center studied the behaviour of chimps given food by humans to trade for other food. The chimps did trade food they liked, such as carrots, for preferred food like grapes. But they didn't spontaneously barter these physical items – they had to be encouraged to do so. And because they don't naturally store objects, this sort of behaviour is unlikely to be seen in the wild. Nevertheless, chimps have been found to exchange services, such as grooming, as well as engaging in the oldest profession in the world.

Prostitution, or 'nuptial gift-giving' as the animal behaviourists endearingly call it, is fairly common in insects – but also in apes. A study on wild chimps in the Ivory Coast's Tai National Park found that they exchange meat for sex. Females engaged in sex more often with males who shared their food with them at least once, while stingy males were out of favour.

Archaeologists know that our ancestor *Homo habilis*, which evolved about 2.3 million years ago, ate meat. Some scientists think it unlikely these males realised they could barter meat for sex and that it wasn't until *Homo sapiens* evolved around 200,000 years ago that males finally realised their chances of sexual success improved if they were a good hunter. If modern-day chimps trade meat for sex, it seems surprising that our ancestors wouldn't have done the same.

Determining when bartering started isn't easy, and no-one knows when trading one object for another object began. The advent of money, however, is easier to trace.

Money Matters

A demand for smaller items to trade than whole cows or bales of hay meant that people started bartering everything from shells to animal furs to coconuts – and even human skulls. Lumps of rare metals began to be used for trading about 6,000 years ago, and around 700 BC the Chinese are thought to have cast so-called '*bubi*' out of bronze in the shape of miniature farming tools such as knives and spades.

Metal coins developed about 2,500 years ago. Ancient Lydia, which is now eastern Turkey, stamped coins from around 650 BC. These 'slaters' were made from gold or silver, or a mixture of the two. Their value depended on the ratio of gold/silver from which they were struck.

Paper money didn't materialise for many years. The Chinese gradually started using credit slips to pay taxes on goods such as tea and salt, and they introduced official bank notes in the late 1100s. But, as *The Travels of Marco Polo* reports, these paper notes often became so tattered that people actually threw them away.

Credit Cards

In the early 1900s, US businesses started to realise the benefit of letting loyal customers pay later on, so they developed store cards, which were mainly used for hotel bills and petrol. Gradually, stores began to club together and accept other stores' cards.

The Charga-Plate, created in 1929, was the predecessor of the credit card. About the size of an iPod Nano, this sheet of metal was embossed with the customer's name, city and state, and held a small paper card for a signature. When a purchase was made, the plate was laid in a recess in the ink-covered imprinter and a paper charge slip was pressed down on top.

The idea of a customer paying different businesses using the same card was introduced in 1950 by Frank McNamara and Ralph Schneider, who founded the Diners Club. American Express then created a global credit card network in 1958 for its US customers, with Barclaycard launching the first credit card in the UK in 1966.

Faking It

Holograms on credit cards make it difficult to produce fakes, because to make them lasers, beam splitters, mirrors, lenses and a special film are needed. In 1960, the magnetic stripe was introduced for 'swiping' a card through a card reader. Microchips on cards added even more security, with card readers requiring a PIN to be entered.

There's also big money to be made from forging actual money – and the practice has been around for centuries. Before paper money existed, fraudsters plated less valuable metal coins with gold or silver. Another trick was to shave the edges of coins in a process known as 'clipping', and then amalgamate the shavings into a counterfeit coin. To identify a fake coin, it could be weighed – or melted, as each metal has its own unique temperature at which it melts. To counter fakes, coin edges are now marked with grooves to show if metal has been scraped off, and bank notes often have metallic security threads, holograms and luminescent ink.

In medieval times, forgers were burnt at the stake. In Germany they were boiled alive in oil and in Russia they had molten lead poured down their throat. Today, they will get away with just a hefty fine and a lengthy prison sentence!

"Money is the root of all evil", supposedly. But while it does corrupt, without money to trade and invest, the world as we know it would be a very different place.

The Natural World

ROUND EARTH

Dispelling the dogma
that our planet is flat

Nominated by **Sir Paul Nurse**, Nobel laureate
and president of the Royal Society

A world held aloft by a giant turtle or tortoise might sound ridiculous to
some, but the idea exists in Hindu, Chinese and even Native American
mythology. Indeed, what may now seem like truly bizarre ideas about
the nature of our planet have existed throughout history in many
different cultures. For centuries, Mediterranean and Mesopotamian
societies believed the Earth was a coin-shaped disc upon which sat
the Mediterranean Sea at the centre of a land mass, surrounded by
an ocean that stretched right to the edges of the world. It wasn't until
around 500 BC that the idea of a
flat Earth was questioned by the
Greek philosopher Pythagoras.

"The discovery of the Earth being round is so important, because it teaches us to be sceptical of what we think we know."
Paul Nurse

While much of what has
been written about him is pure
speculation, Pythagoras is
known to have set up a religious

school at Croton, in modern-day Calabria in Italy. This somewhat secretive brotherhood attracted thousands of followers, who studied Pythagoras's mystical philosophy and lived by a strict and slightly odd set of rules, living as vegetarians, wandering around barefoot, wearing specific clothing, as well as bearing a secret symbol – a pentagon around a five-pointed star.

Probably best known for the mathematical theorem that bears his name, Pythagoras was also the first person to suggest that the Earth might be round. He had no hard evidence to support his theory, but simply reasoned that the gods would have created the world as a sphere, as that was the most logical and beautiful shape. A century or so later, Plato supported this idea, claiming that there must be an opposing land mass on the other side of the world to balance us out, but again this was a purely philosophical conjecture.

It fell to one of Plato's students – the Greek philosopher Aristotle – to come up with scientific evidence that the Earth was round. "At that time, anyone with common sense could see that the Earth was flat," says Nurse. "But Aristotle knew better. He observed that star constellations rose in the sky as a traveller journeyed south, and that the Earth's shadow on the Moon during a lunar eclipse was circular. Both these facts demonstrated that in fact the Earth is round."

By around 330 BC, most people had accepted this as fact. But another question remained: how big was the Earth? In his work *De Caelo*, or 'On the Heavens', Aristotle had deduced that our planet was smaller than some other celestial bodies: "All of which goes to show not only that the Earth is circular in shape, but also that it is a sphere of no great size: for otherwise the effect of so slight a change of place would not be quickly apparent."

Scholars spent years trying to work out the size of the Earth. Eventually, around 240 BC, the Greek mathematician Eratosthenes cracked it. Without setting foot outside Egypt, he calculated that the circumference of the planet was 250,000 stadia (40,555 km). Incredibly, this figure is less than 1 per cent off today's accepted figure of 40,072 km.

CALCULATING THE EARTH'S SIZE

Eratosthenes worked out the size of the Earth using shadows cast in the Egyptian cities of Alexandria and Syene (modern-day Aswan), which were on the same longitude. On the summer solstice, there was no shadow in a well in Syene, where the Sun was directly overhead, but there was a shadow cast by a tower in Alexandria. As he knew the height of the tower and the length of the shadow, he could work out the angle between the two points. He realised this was the same as the angle between Alexandria and Syene from the centre of the Earth – which he calculated to be 1/50th of 360°. Multiplying the distance between Alexandria and Syene by 50, he calculated Earth's circumference to be 250,000 stadia.

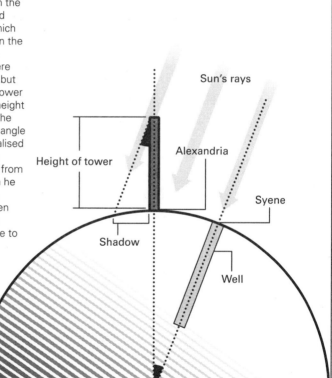

Sun's rays

Height of tower

Alexandria

Syene

Shadow

Well

MAPS

The key to exploring our world

Nominated by **Dermot Caulfield**, editor of
BBC science show *Bang Goes the Theory*

A rare mammoth's tusk was uncovered at Mezhirich, in modern-day
Ukraine, in 1966. While the discovery of such a tusk would be interesting
but not particularly exciting, what made this find so special was the fact
that inscribed upon it was the oldest known map, dating from around
12,000 years ago. This could be trumped by one found on a clay tablet
near Kirkuk, in modern-day Iraq, which some claim to be 27,000 years
old. However, the verdict is still out on this one. What is certain, though,
is that for thousands of years people have needed a written guide as they
ventured further and further from home.

The Greek philosopher Anaximander, who lived from around 610
to 547 BC, could be considered the first scientific geographer and
cartographer. His 'map of the world' only featured the lands that he
knew, and depicted a cylindrical world with a flat top. The idea of a
flat Earth was accepted as fact, until around 500 BC, when Pythagoras
suggested it might be round (see page 51). A few centuries passed
before Aristotle proved this to be the case by observing that the Earth's

NAVIGATING THE WORLD

Today many people have lost the ability to navigate by anything other than Google Maps on their smartphone or the slightly grating tones of their satnav, but for thousands of years mankind relied on the Sun and stars for guidance.

By around 150 BC, a device known as an 'astrolabe' had been invented to work out location. If a traveller knew the time, he could calculate the position of the Sun and stars using the device. This would indicate his latitude, or how far north or south of the equator he was. Conversely, if he knew the latitude, he could use the astrolabe to work out the time.

The magnetic compass was invented ancient China, but not used for navigation until around AD 1100 (see page 157). Even with this device, sailors still struggled to work out longitude, or how far east or west they were. They had to resort to sailing north or south to the latitude of their final destination, then head directly east or west and hope eventually to hit land. Fortunately, in 1759, English clockmaker John Harrison devised the 'chronometer' – a super-accurate and portable clock that could be compared with another clock set to local time. By calculating the difference in time, sailors were then able to work out their longitude.

shadow on the Moon was curved and that stars spotted by travellers on their journey south couldn't be seen in the northern hemisphere.

In the twelfth century, the Arab geographer and cartographer Muhammad al-Idrisi collated information from merchants travelling all over the world to produce the most accurate map of the time, which he showcased to the King of Sicily in 1154. Almost 400 years later, Flemish cartographers Gerardus Mercator, Gemma Frisius and Gaspar Myrica created the first known world globe in 1536. Mercator was the key cartographer of his time, producing detailed maps of everywhere from the Middle East to northern Europe; he was the first to map the British Isles. He also reportedly encouraged Abraham Ortelius to compile the first modern world atlas. Mercator's groundbreaking work led to a lunar crater being named after him.

In 1645, around half a century after Mercator died, Belgian astronomer Michael Florent van Langren mapped the Moon for the first time. In the centuries that have followed, maps of the world outside our world have become ever-more detailed – and, today, you can even explore the mountains and canyons on the surface of the Red Planet using Google Mars.

CONTINENTAL DRIFT

How map-makers and meteorologists moved the world

Nominated by **Professor Bill McGuire**, Benfield Professor of Geophysical Hazards and Director of the Benfield Hazard Research Centre at University College London

Strip off the oceans and, from space, the Earth's surface looks a bit like a football with uneven-sized panels stitched together. These panels are known as 'tectonic plates'. They float around on a viscous layer, called the 'asthenosphere', driven by hot convective currents deep inside the Earth's mantle, in a process known as 'continental drift'.

When two of these enormous lumps of rock collide at what's called a 'subduction zone', one gets shoved under another and forced down towards the mantle far below. As the convergent plates grind together, immense pressure builds up. Just as a rubber band snaps when stretched too far, when the pressure gets too much, the rock suddenly fractures, causing earthquake tremors.

If the source of the earthquake – the epicentre – is shallow and near a city, it can topple tower blocks, flatten houses and twist bridges, causing mass devastation. But, when an earthquake happens on the seafloor, it can be just as destructive – even when located thousands of kilometres from the epicentre.

On Boxing Day 2004, enormous waves, up to 30 m tall, engulfed shorelines around the Indian Ocean. The cause: a 'megathrust earthquake'. Registering 9.3 on the Moment Magnitude Scale, it was the third largest earthquake ever recorded, causing the Earth's crust to rupture in not one but two sections, displacing 30 km^3 of water and creating a series of waves so powerful they travelled round the world for 40 hours.

Understanding how plate tectonics causes earthquakes and the theory of continental drift has taken many centuries to evolve. The first inklings of a malleable, changing planet dates back to the sixteenth century.

The King's Map-maker

Way back in 1596, Flemish cartographer and geographer Abraham Ortelius was the first to propose the radical idea that the earth beneath his feet could be moving. As he travelled the globe, making maps for the likes of the King of Spain, Philip II, he noticed that the east coast of South America and the west coast of Africa could fit together almost perfectly, like a jigsaw if they were just a bit closer – around 5,000 km closer.

Ortelius had realised that whole continents were shifting: "torn away from Europe and Africa ... by earthquakes and floods ... The vestiges of the rupture reveal themselves, if someone brings forward a map of the world and considers carefully the coasts of the three [continents]." As the creator of the first modern atlas, the *Theatrum Orbis Terrarum*, Ortelius was as well qualified as anyone to propose such a radical idea. He was just a few centuries too early.

The idea of drifting continents festered in the backwaters of science until, in 1912, a German meteorologist called Alfred Wegener recognised its importance. If Ortelius was right and continents such as Africa and South America had been joined together, then all the continents could once have been one great mass of land. A land mass that Wegener called '*Urkontinent*'.

"Wegener's idea of drifting continents was not Earth-shattering. But it was Earth-building: seeking as it did to begin to construct a model that explained how our planet worked," says McGuire.

Wegener wasn't actually the first to come up with this idea. In 1889, Italian geologist Roberto Mantovani suggested that one large supercontinent once existed. But it was Wegener who went a step further, presenting the concept of continental drift at the Greenough Club – the undergraduate geological society at University College London. Not being a geologist put Wegener at a disadvantage, though, and with no firm idea of the exact mechanism of how continental drift worked, he was met with scepticism.

"A vote followed his talk, and the idea was soundly defeated," says McGuire. "Not far off a century later, the same lecture room is used to teach Earth Science students; all are familiar with plate tectonics and supercontinents. But, sadly, I suspect that for many the name Wegener means nothing."

A Lonely Death

It was during his many expeditions to Greenland that Wegener formed his theory of continental drift. But it was one of these trips that was to be the end of him. In the winter of 1930, snow hid route markers back to civilisation, making the return journey perilous. Hunger, fatigue and temperatures of –60°C conspired against Wegener's group. A penknife

SUPERCONTINENTS

Ever looked at your fingernails and wondered how fast they grow? On average, it's about 10 cm a year – the same speed that most tectonic plates move.

Admittedly, 10 cm a year doesn't sound like much, but, over the aeons of time, continents can shift vast distances. If we could travel back 200 million years and launch a satellite into space to orbit the Earth, the photographs of our planet would look a very different from those today. During the Permian period, the continents were all bundled together in a vast supercontinent known as 'Pangaea' (Wegener's *Urkontinent*), surrounded by one enormous stretch of ocean called 'Panthalassa'. How do we know this? The fossil record.

In the late Permian and early Triassic periods, there existed a large lizard-like creature that was the size of a dog, called *Lystrosaurus*. A herbivore, *Lystrosaurus* was an incredibly common beast, making up 95 per cent of species found in the fossil record from that time and roaming over much of the terrestrial world, from Antarctica to India to Russia to China. With a somewhat chunky build, *Lystrosaurus* wasn't made for long-distance swimming, so must have spread to the four corners of the planet on foot. To do that those far corners had to be connected by land – one big hunk of land, Pangaea.

Fifty million years later, during the Triassic period, Pangaea split apart into Laurasia, which is modern-day Europe, Asia and North America, and Gondwanaland, today's Antarctica, Australia, Africa and South America. In 250 million years' time, all the continents will have regrouped into another supercontinent, which scientists are calling 'Pangea Proxima'. One wonders, by then, how many times Wegener's remains will have circumnavigated the globe.

claimed the frost-bitten toes of one of the group, Fritz Loewe. But worse was to come. As food grew more scarce, the group decided to split up. Wegener and another colleague, Rasmus Villumsen, ventured on towards the next camp.

But they never made it. Wegener's body was found six months later, buried by Villumsen, who himself was never seen again. The irony is that Wegener's final resting place now lies 2 m further from his homeland, due to continental drift. Sadly, he died not knowing that his theory would one day be a staple of geological textbooks and the progenitor of the model of plate tectonics, which underpins modern geology.

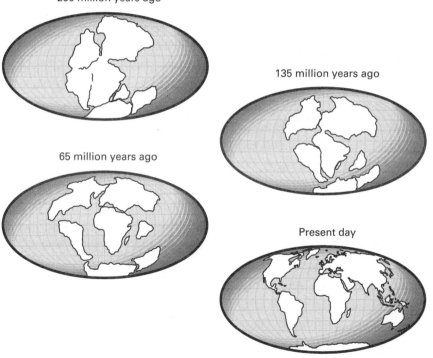

250 million years ago

135 million years ago

65 million years ago

Present day

THE AGE OF THE EARTH

The fossil hunters and radiometric daters
who tracked down the true age of our planet

Nominated by **Stephen Baxter**, best-selling science fiction
writer and author of *Revolutions in the Earth: James Hutton*
and the *True Age of the World*

The oldest person alive today is 114. Tortoises can live for over 150 years, a clam was found in the freezing waters off Iceland in 2007 that was over 400 years old and there's a tree in Sweden believed to be almost 10,000 years old. But all this is a drop in the ocean compared to how long Earth has been around – an incredible 4.5 billion years.

The planets in our Solar System formed when gas and dust swirling around the young Sun smashed together, congealing into bodies 10 km across, known as 'planetisimals'. As these collided they formed large planets, and eventually settled into their orbits around the Sun. But it was many hundreds of millions of years before even very simple single-celled organisms started to develop on Earth, 3.8 billion years ago. And modern humans like us didn't evolve until around 200,000 years ago. That means that if the whole of Earth's existence is compressed into 24 hours then we humans would have existed for just the final four seconds.

If we struggle to get our heads round that vast timescale, imagine what it was like for our ancestors. With nothing much better to go on at the time, in the 1600s an Irish bishop, James Ussher, turned to the Bible for inspiration on ageing the Earth. By adding up the ages of all the people mentioned in the Bible, he settled on the very precise date of 23 October 4004 BC. Ridiculous as this now sounds, Ussher wasn't the only one who put the age of the Earth in the thousands. The normally brilliant mathematician Newton had hit on 4000 BC, while astronomer Johannes Kepler worked out the Earth had formed in 3992 BC.

Fossil Hunters

Ignoring what he read in books, and instead trusting in what he saw with his own eyes, the seventeenth-century Danish biologist and geologist Nicholas Steno travelled all over Europe. He spent much of his time analysing the human body, noting things like how muscles contracted. During his travels, Steno came into contact with numerous scientists, challenging even the likes of Descartes and his theory as to the origin of tears. But it was Steno's observations of fossils that relaunched the enquiry into the age of the Earth.

In 1666, two Italian fisherman landed an enormous catch – a whopper of a shark off the coast of Livorno. The Grand Duke of Tuscany commanded that the beast's head be lopped off and sent to Steno. The Dane's dissection revealed a distinct resemblance between the shark's teeth and what in those days were known as 'tongue stones', strange imprinted shapes that were thought to grow naturally within rocks.

These 'strange imprints' we now call fossils, literally meaning 'dug from the ground'. Fossils can be formed in a variety of ways: for example, when a marine organism dies, it sinks to the sea floor and the organic parts of the organism are fed on by scavengers and decomposed by bacteria. What remains are the tougher, inorganic parts, made from minerals such as calcium. Over time, as this structure becomes covered by sediment, water containing minerals penetrates into the structure, which then hardens, a bit like a sponge filling with glue.

Steno was on the right track when, alongside contemporaries including Robert Hooke, he realised that fossils were the remnants of living creatures that once roamed the planet. He worked out that the different layers of rock must have hardened from sediment deposited at different times, and so the fossils in the separate layers were of creatures from different eras.

The Father of Modern Geology

People were getting suspicious that the Earth had to be a lot older than originally thought. Then along came a Scot, by the name of James Hutton. "Hutton was the true father of modern geology," says Baxter. "He produced the evidence that Earth had to be vastly older than a few thousand years – it had to be millions, at least."

Hutton started out with an interest in mathematics and chemistry, but, after inheriting a farm outside Edinburgh, his interest grew in geology. As he paced the land, he began to realise that the soil beneath his feet was made from broken rocks. "It started to dawn on Hutton that, as well as the destructive force of erosion, there had to be creative forces in the Earth as well," says Baxter. "So he went hunting for evidence of uplift."

Hutton discovered intrusions of granite (younger rock layers inside older rock), but also, crucially, unconformities. These are where strata of rock are laid down, uplifted and torn apart, with new strata being laid on top – a bit like a book lying flat on another one held upright. "The unconformities were evidence that the Earth doesn't just erode gradually but goes through cycles of uplift and erosion," says Baxter. "But the intrusions were evidence of heat within the Earth."

A natural engine that drove molten rock up from below the Earth's surface was a radical idea, but nonetheless Hutton could see no other explanation than the planet having a hot interior. The idea was called 'plutonism', and elbowed out the competing theory that all rocks had precipitated out of water in one gigantic flood, which supported the religious belief in Noah's flood.

"James Watt, who developed the steam engine, was a good friend of Hutton's," says Baxter. "Hutton came up with the idea of the Earth being like a huge, hot engine. He saw uplift as proof of what we now call the rock cycle."

Despite being immersed in the world of the Royal Society in Edinburgh, Hutton wasn't a man to seek recognition for his theories. After a couple of decades of pontificating, he did eventually start writing about his ideas. But the three hefty volumes of over 2,000 pages certainly didn't make for easy reading. "Hutton is often forgotten for his contributions to science, not only because he was a lousy writer but also because he couched his theory in religious terms, which was falling out of fashion at the time. I like Hutton because he was very down to earth. He just opened his eyes and walked around Britain, observing and extrapolating what he saw in the ground into a fundamentally correct vision of the Earth, its geological processes and its age."

So what figure did Hutton come up with for the age of the Earth? He realised that the process of seabeds uplifting into mountain ranges was very gradual, requiring vast amounts of time – much, much longer than a mere few thousand years. But what he didn't quite get right was his theory of 'uniformitarianism', which claims that the only forces that have ever shaped the world are the ones we see now. This idea excludes cataclysmic events, such as floods.

Fellow Scot Charles Lyell used the idea to form the basis of modern geological thinking in his book *Principles of Geology*, which was picked up by the young Charles Darwin, even influencing his thoughts on his theory of natural selection.

Neither Hutton nor Lyell ever made a stab at an exact figure for the age of the Earth, but their theories fuelled the debate for others who had already started to think in terms of millions of years.

Hot Earth

In the 1790s, British naturalist William Smith had worked out that if two layers of rock in entirely separate places had the same types of fossils then it was likely that these layers were the same age. Working on that assumption in the early 1800s, Smith's nephew, John Phillips, calculated that the Earth could be 96 million years old. Still a bit far off the mark, but getting closer.

Then along came William Thomson, better known as Lord Kelvin. Born in Belfast in 1824, he survived a heart condition which almost killed him aged nine, before starting to study at Glasgow University at the tender age of ten, which was apparently quite normal for bright sparks in those days. He went on to have a pretty impressive career as a physicist and engineer, from coining the term 'kinetic energy' to having a unit of temperature measurement named after him. Kelvin found fame and fortune as an inventor and engineer of electric telegraphs, becoming the first UK scientist to be a member of the House of Lords, after he was knighted by Queen Victoria for his work on the transatlantic telegraph project.

It is Kelvin's work on ageing the Earth that we're interested in here. Despite being a Christian who loosely believed in creationism, he reckoned that some dynamic process had kick-started the birth of the Universe, followed by a gradual cooling, where life on Earth could develop. Although Kelvin's idea supported uniformitarianism, where changes in the Earth's structure were gradual, it contradicted the idea that this process happened at a constant rate.

Back in 1749, geologist and naturalist Georges-Louis Leclerc (who renamed himself le Comte de Buffon after he inherited a mini-fortune) published results from an experiment he'd set up to calculate the cooling rate of a hot Earth. He forged a number of different-sized iron balls and timed how long each took to cool. By extrapolating this to the size of the Earth, he estimated the planet was 75,000 years old. He was way out, but way ahead of his time.

When Kelvin had a go in 1864, using his knowledge of the melting temperature of rocks, he came up with the initial estimate of 20 to 400 million years old. Over time he honed this down to 20–40 million, but didn't take into account that the planet has a fluid mantle, so was still many millions of years out. This figure was challenged throughout his life, but it wasn't until he was on his deathbed, that his estimate was officially kicked out by a major breakthrough in dating rocks.

The Ultimate Date

In 1905, Ernest Rutherford suggested that the recently discovered phenomenon of radioactivity could be exploited to work out the age of extremely old geological samples. At the heart of the idea is the fact that radioactivity is the result of the disintegration of unstable atoms of one chemical element into atoms of another (see page 69). The disintegration takes place at a constant rate characterised by the so-called 'half-life' of the radioactive atom, which gives the time after which just half of the original 'parent' atoms remain, the other half having been turned into so-called 'daughter' atoms of another element. These timescales can be incredibly long – many billions of years – and Rutherford realised that by measuring the relative proportion of parent to daughter atoms in a sample, it would be possible to tell how old the sample was.

By comparing the concentration of a stable isotope in a rock with the initial concentration of the element, its age can be calculated. So, to work out the minimum age of the Earth, scientists just needed to find the oldest rock on the planet.

In 2008, in Hudson Bay, northern Canada, geologists found the oldest-known rock formation in the world. Radiometric dating established that the Nuvvuagittuq greenstone belt was up to 4.28 billion years old. These rocks have seen a lot in their lifetime, and will probably see a lot more, as our planet should stick around for many billions of years to come.

RADIOCARBON DATING

How radioactivity unlocked the true age of fossils

Nominated by **Professor Chris Stringer**, research leader in human origins at the Natural History Museum, London

Alarm bells should have started ringing straight away for those attending the meeting at the Geological Society in London on 18 December 1912. Charles Dawson's account was fairly sketchy of how he'd come to lay hands on the fossilised remains of a previously unknown early human, *Eoanthropus dawsoni* – the infamous 'Piltdown Man'. Dawson claimed that a workman at the Piltdown gravel pit in East Sussex had given him part of a broken skull

"Radiocarbon dating allowed us to properly test the chronologies established by historical and archaeological research for the first time. It has verified the age of important human fossils and artefacts, while demonstrating that other relics such as Piltdown Man and the Shroud of Turin were far less ancient than had been claimed."

Chris Stringer

four years earlier. After showing it to Arthur Smith Woodward, the Geological Department's keeper at the British Museum, Woodward joined him at the site, recovering yet more pieces of the skull and the lower jawbone.

At the Geological Society's meeting, Woodward claimed that Piltdown Man must represent the missing link between ape and man. While it was accepted by many in the scientific world, there were some who challenged the theory right from the start. In 1913, David Waterston from King's College London claimed it was no more than an ape's jawbone and a human skull. Two decades later, anatomist Franz Weidenreich correctly spotted that the jawbone belonged to an orang-utan, but another twenty years were to pass before, in 1953, a chemical dating technique – the fluorine absorption test – proved that Weidenreich had been right.

No-one knows for sure who was responsible for the forgery. Charles Dawson is the obvious culprit (having been accused of falsifying 38 other 'finds'), but nearly every British researcher involved with Piltdown has also been named as a suspect. The advent of radiocarbon dating put the final nail in the coffin of Piltdown Man. And it was the discovery of radioactivity that was the key to the invention of radiocarbon dating.

Discovering Radioactivity

Antoine Henri Becquerel was one of a long line of successful Becquerel scientists, becoming the third in his family to take on the position of chair of physics at the Natural History Museum in Paris. It was while he was at the museum that he made his biggest scientific breakthrough.

The recent discovery of X-rays intrigued Becquerel, as he thought there must be some link between them and the way that certain materials glowed when exposed to bright light. In 1896, he set up an experiment, exposing to light some uranium salts his grandfather had passed on to him, before wrapping them up in a black cloth and placing them between some photographic plates. Returning a while later, he discovered the image of the crystals on the plates, but this also happened when Becquerel didn't expose the uranium to light. Baffled by the strange rays that he realised must be emanating from the compounds, he charged one of his students with finding out what they were. That student was Marie Curie.

Working alongside her husband, Pierre, Marie set about trying to solve the puzzle of these strange rays. They discovered that the mineral ore pitchblende which uranium comes from emitted vast amounts of energy in the form of these rays, but, bizarrely, the pitchblende never lost mass. The Curies also found that thorium, radium and polonium emitted this energy. (The element radium was discovered by Pierre and Marie Curie, and Marie also discovered polonium, which she named after her native Poland; her maiden name was Sklodowska.)

Becquerel and the Curies called this new energy 'radioactivity' and in 1903 they shared the Nobel Prize in physics for their work. Marie was the first woman to receive a Nobel Prize, which was the first of two for her, each in a different field of science. No other person has ever received two Nobel prizes in two different scientific disciplines. The unit of radioactivity – the curie – is named after her, as is the element curium. But Marie was to pay the price for her long hours spent exposed to radioactive materials.

Marie Curie died in 1934 from aplastic anaemia, a condition where the bone marrow doesn't produce enough new blood cells. Her death was almost certainly the result of overexposure to radiation. Even today, her notebooks are so laced with radioactivity that they have to be kept in lead-lined boxes. But her death wasn't in vain, her part in the discovery of radioactivity was the key to opening up a whole new branch of science.

Alpha and Beta

Chemists Ernest Rutherford and Frederick Soddy had identified two types of radioactive particle in 1899, which they named 'alpha' and 'beta'. Alpha particles produce strong ionising radiation but can be stopped by a sheet of paper, while beta particles are lighter and produce weaker ionising radiation but can travel through several millimetres of metal. While the discovery of alpha and beta particles was groundbreaking, possibly the duo's bigger breakthrough was revealing the so-called 'half-life' of radioactive materials.

The unstable atoms in radioactive elements mean that they disintegrate, eventually decaying into the stable atoms of another element. The 'half-life' of the radioactive atom describes the constant rate at which the decay occurs: it gives the time after which half of the original 'parent' atoms remain, the other half having been turned into atoms of the so-called 'daughter' element. Rutherford realised that, by measuring the relative proportion of parent to daughter atoms in a sample, one could tell how old the sample was. As the age was often in the billions of years, by comparing the concentration of the stable atom with the initial concentration of the element, he was able to work out the age of rocks and of the Earth itself (see page 62).

Radiocarbon Dating

The half-life is also a useful tool for dating fossils. When an organism dies, it stops eating plants and so no longer absorbs the isotopes carbon-12 and carbon-14. (Isotopes are atoms of the same chemical element, which differ in the number of neutrons.) Over time, carbon-14, which is radioactive, decays, while carbon-12 in the fossil remains at the same level as it is more stable. By comparing this ratio in fossils to the baseline level in the atmosphere, the age of the fossil can be calculated. In this way, radiocarbon dating proved that the skull and jawbone of Piltdown Man were both recent in age, confirming the findings of the less precise methods that had been used before.

THE THREE-AGE SYSTEM

How the discovery of
stones shaped by
prehistoric man stamped
on the Creationist view

Nominated by **Dr Francis Pryor**, archaeologist,
presenter of *Time Team*, author of books including
The Making of the British Landscape, and co-discoverer
of the Bronze Age site Flag Fen

Controversy raged in the eighteenth century as to whether so-called 'thunder stones' were fossils or had been shaped by man. French antiquarian Nicholas Mahudel was particularly interested in prehistoric objects, and was convinced that the stones had been shaped by man, so he presented his idea to the Académie des Inscriptions et Belles-Lettres in Paris. The problem was that if the stones had indeed been honed by someone using tools, the verdict was that they couldn't possibly be more than a few thousand years old, as it would contradict the biblical version of the Creation.

Despite resistance from some quarters, Mahudel continued with his investigations of burial sites, and drew the conclusion that there had been three stages of development – stone, bronze and iron.

In 1836, Danish museum curator Christian Jürgensen Thomsen took the idea a step further. By classifying artefacts into categories, he spotted a pattern: stone artefacts had been used initially, then came the rise of bronze artefacts with the discovery of how to make bronze implements, before the same happened with iron. So, after hours poring over records of finds and bent over prehistoric artefacts, Thomsen had succeeded in compiling empirical evidence that the three ages had occurred – a long time ago.

The impact on society and religion was immense. As Pryor points out: "Ideas can be brilliant, but they have to be accepted by society at large if they're to have any practical effect. Without Thomsen's three-age system, and the advances in geological theory that went with it, Darwin's revolutionary views on evolution might never have caught on as fast as they did, as they would have met far more resistance in academic and scientific circles."

DID YOU KNOW?

The 'ages' do not refer to an absolute time period. Shifts from one age to another occurred gradually over long periods of time and at different times in different parts of the world. For instance, India and the Middle East had moved from the Bronze into the Iron Age by around 1200 BC, whereas the Iron Age did not begin in Europe until around 600 BC.

THE HAND AXE

The longest-used tool in human history

Nominated by **Dr Dave Musgrove**, Editor,
BBC History Magazine

"The hand axe takes us back to the earliest stirrings of what we can see as a modern human mind. These tools were multipurpose chopping devices that would have been of considerable practical value to early humans, but it's what they tell us about the brains that made them which really matters. You can't make a hand axe without having creative vision and the ability to carry through that vision: human faculties that have stood us in good stead for the intervening 1.5 million years since these tools first started to appear."
Dave Musgrove

Not to be confused with a modern, wood-handled axe, the prehistoric hand axe was shaped from rock, such as flint, to give a sharp edge for cutting meat. Typical of the lower and middle Palaeolithic era, its use stretched from 1.5 million to around 50,000 years ago, making it the longest-used tool in history.

One of the first hand axes brought to our attention was discovered in northern France by geologist Joseph Prestwich and archaeologist John Evans in 1859, the year that Charles Darwin published *On the Origin of Species*. At that time, it was believed that humans had only been around for a few thousand years. When Prestwich and Evans presented the axe at a lecture to the Royal Society and revealed that it was 400,000 years old, it began to be accepted that humans had walked the Earth many thousands of years longer than previously believed.

Since then much older specimens have been uncovered by archaeologists in Africa. The Palaeolithic – an era when early humans banded together in small societies and lived by gathering plants and hunting or scavenging wild animals – was characterised by the development of knapped stone tools. (Palaeolithic literally means 'old stone age'.) The oldest of these artefacts, discovered in Ethiopia, were made 2.6 million years ago. Known as the 'Oldowan tools', they include choppers, scrapers and pounders, and they are considered the precursors to the hand axe.

The hand axe, developed 1.5 million years ago in Africa, involved chipping the stone on both sides to create two cutting edges. The shift from primitive chopper-like tools to these more carefully crafted double-faced axes marked an important milestone in the history of technology, and gave the early humans who wielded them an advantage in the battle to survive.

Most hand axes were made of flint, but other stone, such as quartz, was used too; the rock just had to be hard enough to withstand being chipped into the right shape. There were several basic shapes – oval, triangular, pear-shaped – but there are disagreements about their uses. The American neurophysiologist William H. Calvin has theorised that the rounder shapes could have served as 'killer frisbees' used to stun one animal when thrown at a herd gathered by a waterhole, for example. There is not much support for this theory, though, mainly because it is doubtful that a hand axe would penetrate deeply enough to stun the animal. Generally, it is thought the hand axe was used for cutting meat and extracting bone marrow. Studies conducted at Boxgrove in West Sussex in the 1990s, involving a butcher cutting up a carcass with a hand axe, back this up. The butcher found that the hand axe was ideal for getting to the marrow, which, being full of protein and vitamins, would have been prized food in Palaeolithic times.

Hand axes have mainly been found in Africa, Europe and northern Asia. The oldest European specimens uncovered at two sites in Spain in the 1970s were recently dated using a technique called 'magnetostratigraphy', which is based on the periodic reversal of the Earth's magnetic field, and found to have been crafted between 760,000 and 900,000 years ago.

Hand axes continued to be used, largely unchanged, until around 50,000 years ago, when many new tool types appear in the archaeological record – projectile points, engraving tools, knife blades and tools for drilling and piercing. This shift in tool technology, known as the 'Upper Palaeolithic Revolution', led to a population explosion of modern humans. It's possible that this, as well as the dramatic variation in Europe's climate conditions from cold to warm, may have caused the demise of the Neanderthals. Less agile than humans, the Neanderthals' ambush strategy may have taken a hit as the forests receded and their hunting grounds shrank, eventually leading to their extinction.

GAIA THEORY

How searching for life on Mars
led us to protect our own planet

Nominated by **Fred Pearce**, author whose books
include **Peoplequake** and **When the Rivers Run Dry**

"We learned a lot about the Moon, but what we really learned was about the Earth. From the Moon you can put your thumb up and hide the Earth behind it. How insignificant we really are, but then how fortunate we are to enjoy living here amongst the beauty of the Earth."

"Gaia has been immensely useful and democratising, because it's helped a lot more people become environmentally aware of this fragile planet."
Fred Pearce

Sentimental words from *Apollo* astronaut Jim Lovell, during an interview in 2007 for the film *In the Shadow of the Moon*. But while our planet feels monumental to us down here on Earth, when astronauts take a step back and look at it immersed in the vastness of space, they must be in awe of its ability to survive the vast, harzardous Universe.

Maybe it's not surprising then that it was a scientist, who formerly worked for NASA, who first realised how the Earth must be able to regulate itself in order to survive the punishing extremes of space.

British scientist James Lovelock had never been into space. But in the 1960s he joined a NASA team which was designing instruments for lunar probes and searching for life on Mars. But getting to the Moon was costly, let alone planning a future round trip to Mars. Lovelock began to ponder whether there could be another way to reveal the secrets of the Red Planet.

Having returned to the UK in 1963, he became an independent science consultant. Once, when visiting the Jet Propulsion Lab in California, it suddenly dawned on him that we didn't need to go all the way to Mars to look for signs of life. Just as the Earth's atmosphere is chemically active, using an infrared telescope he could see whether the Red Planet's atmosphere contained carbon dioxide. If so, it would be a tell-tale sign that life exists.

Astronomers started looking for signs of carbon dioxide in Mars's atmosphere – and found it. While probes and the Mars rovers that have since been to the Red Planet have yet to find any form of life, what Lovelock's discovery did achieve was to open his mind to the idea of the Earth itself being a 'living planet'.

In 1979, Lovelock published a controversial book on the idea of the Earth being like one giant self-regulating superorganism, where all the planet's elements work together as a complex, integrated system to maintain a climate and biology that is just right for life, just as feedback mechanisms maintain the human body at a relatively constant 37°C. Similarly, through geological and chemical processes, the air we breathe at sea level remains at about 20.8 per cent oxygen.

"Life is managing the planet to keep the environment fit for life," says Pearce. "For instance, the Earth maintains an atmosphere that is the right temperature and doesn't have too much oxygen that the whole place goes up in flames."

Daisyworld

Lovelock called his idea 'Gaia theory'. But it began to draw extensive criticism, maybe because of its slightly whimsical name, after the mythical Greek goddess of the Earth, which had been suggested by Lovelock's neighbour, literature Nobel prizewinner William Golding, or possibly because some other scientists didn't feel it sat comfortably with other theories.

"Richard Dawkins had a problem with the idea of natural selection working at that kind of *communal* level and going against the selfish gene idea," says Pearce. The 'selfish gene' theory suggests that if two individuals are genetically similar then they're more likely to behave selflessly, which explains altruism – when an organism helps another to its own detriment. "But look at how bees behave at a communal level in a hive," says Pearce. "They operate cooperatively, because it is in their interest to do so."

Indeed, Lovelock retaliated by conjuring up another idea – 'Daisyworld'. On this metaphorical planet, there exists two types of daisies, black ones that bloom in the cold and white ones which flourish when it's hot. But as black daisies absorb heat and white ones reflect it, for the planet to keep functioning, it makes sense for it to stabilise at a warm temperature in which both types of flowers can exist.

Green Earth

Despite the critics, Gaia theory has done for the environmental movement what no other concept could have. "Environmental thinking used to be very compartmentalised – lots of people with different ideas and enthusiasms, wanting to protect different species and habitats. Gaia theory is an idea that integrates a lot of environmental concerns in one image of the planet," says Pearce. "The planet is fit for life, not just by chance, but because through evolution life forms have kept it that way. That gives us a new perspective on the environmental damage that we are doing to the planet. By emitting carbon dioxide which has been buried underground by nature for tens of millions of years, we are upsetting those processes, we are pulling at Gaia's control mechanisms."

Rather than bury their heads in their hands and give up in despair, many scientists, such as climatologists and meteorologists, have found Gaia theory to be important in helping them frame what to research. For example, the Gaian idea of feedback mechanisms maintaining the Earth's equilibrium was applied when researching the effect of marine algae on cloud generation.

"Algae in the oceans generate dimethyl sulphide, which is a powerful cloud condensation nuclei creator," says Pearce. "When there's lots of algae, lots of dimethyl sulphide creates more clouds, which have a cooling effect. That feedback loop stabilises the atmosphere."

THE GREENHOUSE EFFECT

The heat trap that's vital for life on Earth

Nominated by **Dr David Adam**, op-ed editor of *Nature*

French mathematician and physicist Jean Baptiste Joseph Fourier guaranteed his name in the annals of science with his complex, yet hugely influential, mathematical operation known as the 'Fourier Transform' (see page 316).

But he also played a hand in the discovery of one of today's hottest topics – climate change. Fourier is credited with coming up with the idea of the 'greenhouse effect'. He realised that the Sun was too far from the Earth to keep the planet at its relatively balmy temperature. So he worked out that there must be some other mechanism heating the Earth and suggested that the atmosphere could be trapping heat.

The Frenchman didn't call his discovery the 'greenhouse effect', but future scientists named it that after an experiment by a Swiss aristocrat which influenced Fourier's work. Through his experiments, Horace-Benedict de Saussure had found that the temperature inside a cork was higher if small panes of glass were inserted into it. Fourier deduced that if the atmosphere worked a bit like the glass panes, heat would be trapped inside. In reality, it's not a great analogy.

"The greenhouse effect is probably the most important badly named idea in science," says Adam. The way the greenhouse stops your tomatoes getting chilly is completely different to how the Earth stays warm. The glass or plastic coverings in greenhouses stop warm air, and therefore also heat, escaping by convection. In contrast, the greenhouse effect relies on gases in the Earth's atmosphere absorbing and then re-emitting thermal radiation.

Adam explains: "The Sun emits energy as visible light, which is absorbed at the Earth's surface, and re-radiated as thermal radiation. Some of this heat is absorbed by greenhouse gases in the atmosphere and returned to Earth, making it warmer than it would otherwise be."

The 'greenhouse gases', such as carbon dioxide and methane, are vital to life on Earth. "The greenhouse effect is commonly portrayed as a menace," says Adam. "But without it, our planet would be well below freezing. The greenhouse effect really did change the world, from a frozen and barren dead rock to a thriving mass of life."

The reason the greenhouse effect has attracted such bad press is its association with the wrong gang: global warming. "The problem of man-made global warming is caused by humans adding to the heat-trapping gases by burning fossil fuels," says Adam. "This 'enhanced greenhouse effect' is the problem."

HOW THE GREENHOUSE EFFECT WORKS

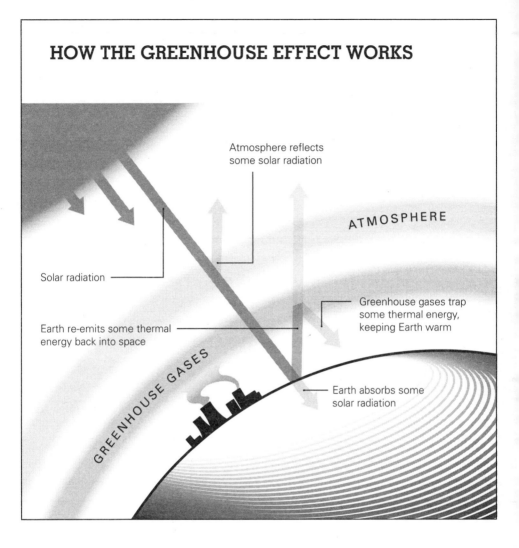

Atmosphere reflects some solar radiation

ATMOSPHERE

Solar radiation

Greenhouse gases trap some thermal energy, keeping Earth warm

Earth re-emits some thermal energy back into space

GREENHOUSE GASES

Earth absorbs some solar radiation

The Universe

THE TELESCOPE

The invention that opened our eyes to the heavens

Nominated by **Dr Stuart Clark**, author whose books include *The Big Questions: The Universe* and *The Sun Kings*

Born in 1564 in the leaning-tower town of Pisa, Galileo Galilei considered joining the clergy but ended up becoming a mathematician. His work covered scientific fields as diverse as the nature of comets and the tides. One of his major contributions to science came in 1609 when he appropriated an idea from a Flemish spectacle-maker.

"The telescope allowed us to leap beyond the naked ability of the human senses. By showing us that there were objects in the sky hidden from normal human view, it spurred not only the scientific revolution but also our cultural perception of the world. It made us re-evaluate our place within the Universe."
Stuart Clark

Spectacles with concave or convex lenses had been around for centuries. But Hans Lippershey had the ingenious idea of putting a concave lens at one end of a tube and a convex lens at the other. When he peered through one end of his 'spyglass', objects appeared larger than in real life, magnified to around three times their true size. The telescope was born.

When Galileo heard about this device, he realised its massive military potential. If Ottoman Turk invasions could be spotted from miles away, there would be more time to prepare for attacks. In the summer of 1609, he set about building his own version. Once he'd managed to improve it to eight times magnification, he showed it to the Doge of Venice, who was so impressed that he offered Galileo a lifetime of security with a fixed salary, but also a fixed address – in Venice.

Galileo turned the offer down, as he wanted flexibility to practise his science. He'd found another use for his telescope. When he turned it towards the heavens, its magnification let him view sights that no-one had seen before, like the rings of Saturn, the four largest moons of Jupiter and strange shapes on our own lunar neighbour.

Galileo could see that the Moon wasn't a perfect sphere at all, but pockmarked with craters that we now know were caused by asteroid collisions. At that time, people believed that only things that existed on Earth were changeable or damaged. The idea that heavenly bodies could be non-spherical and imperfect was blasphemous.

DID YOU KNOW?

Launched in 1990, the Hubble Space Telescope is the only telescope designed to be serviced in space by astronauts. This is just as well. Famously, it turned out that the primary mirror had been ground to the wrong shape – something that had to be fixed by crew members of the Space Shuttle *Endeavour* in December 1993.

It turns out that Galileo was not actually the first to observe the Moon in such detail. English astronomer Thomas Harriot wrote a diary containing maps of the Moon, dating from a couple of months before Galileo published his own sketches. But Galileo was ever the publicist and networker. This stood him in good stead for having many of his ideas accepted, and even gaining a successful audience with the Pope in 1611. But, in the end, he pushed the Church too far.

First, the fiery, stubborn Italian managed to wind up some Jesuit astronomers. Then he became more vocal about his support for Copernicus's heliocentric model of the Solar System (see page 95). This was to be his downfall. The Inquisition leapt at the chance to charge Galileo with breaking the ban on defending Copernicanism – and he ended his days under house arrest.

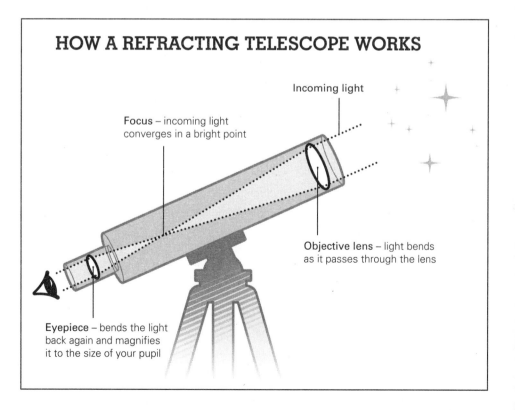

HOW A REFRACTING TELESCOPE WORKS

Incoming light

Focus – incoming light converges in a bright point

Objective lens – light bends as it passes through the lens

Eyepiece – bends the light back again and magnifies it to the size of your pupil

THE HUBBLE SPACE TELESCOPE

The eye in the sky

Nominated by **Dr Caroline Smith**, Curator of Meteorites at the Natural History Museum, London

"For over two decades, Hubble has provided us with a wealth of information about our Universe and all the amazing objects it contains. Among its many other achievements, it has allowed us to peer into the far distance of space and time to see planets forming around young stars, the same sorts of processes which were happening in our Solar System 4.6 billion years ago."
Caroline Smith

Stars 'twinkle' because of atmospheric distortion caused by air pockets that exist between space and the Earth. Combined with the fact that the atmosphere can block certain types of light from space, ground-based telescopes are limited in what they can capture.

To get round these problems, rocket scientist Hermann Oberth suggested in 1923 launching a telescope into space. This idea was way ahead of technology, and it was seven decades before the world's first optical space telescope was launched, on 24 April 1990.

The size of a large school bus, Hubble orbits 570 km above Earth, travelling at the incredible speed of 8 km per second. It takes just over an hour and a half to complete an orbit of the planet and each orbit uses about the same energy as powering 28 light bulbs for 96 minutes.

Hubble contains what is known as a 'Cassegrain reflector', a combination of a primary concave mirror and a secondary convex mirror. Light hits a primary mirror, then encounters the secondary mirror, which focuses the light through a hole in the primary mirror, which then goes to the telescope's instruments. The telescope needs to stay incredibly steady when taking pictures, so it's been designed to lock on to a target that would be the equivalent of shooting at a human hair from 1.6 km away.

The instruments on board include the wide-field camera, which can see ultraviolet, infrared and visible light, and is used to study dark energy and dark matter (see page 108); a special spectrograph, which searches for black holes (see page 192); and an advanced camera that studies the oldest objects in space from 700 million years after the Big Bang.

Hubble hasn't been without its hitches, though. Straight after launch in 1990, NASA discovered the images Hubble was sending back were blurred, because of a flaw in the mirror. Fortunately, one of the engineers worked out how to fix the problem. He came up with the solution while in the shower. As he swivelled the shower head, he realised he could design a similar device that would tilt Hubble's mirrors to sharpen the image hitting the camera. In 1999, four of its six gyroscopes failed, leaving it with no pointing system, so a Shuttle repair mission had to be launched for the first time over the Christmas period. In its lifetime, Hubble has beamed back hundreds of thousands of images, which have informed over 6,000 scientific papers. Every week, it sends enough data that would fill books stacked on a 1 km-long bookshelf.

Hubble hasn't been cheap – costing $10 billion during its lifetime – but not only has it captured some of the most memorable images of the cosmos, it has helped confirm the existence of so-called 'dark energy', identify objects such as quasars (the intensely luminous centres of distant galaxies) and confirm the age of the Universe.

A CENTRAL SUN

The heliocentric model of the Solar System

Nominated by **Professor Jim Al-Khalili**, theoretical physicist, author and presenter

It was dangerous to be a scientist before the seventeenth century. One radical idea too many and you ran the risk of being locked up, or, worse, roasted alive. Religion has sculpted much of scientific history., nowhere perhaps more so than in the field of astronomy and Earth's position in the Solar System. To question God and his 'grand design' was to commit heresy – and, for that, a bonfire and a stake often awaited you.

"A number of astronomers throughout history suggested that the Earth goes round the Sun, and not the other way round. But it wasn't until Galileo pointed the newly invented telescope up to the sky that the truth became evident."
Jim Al-Khalili

Take the fate of Italian friar and astronomer Giordano Bruno. In the late 1500s, he had the audacity to suggest that more than one world might exist out there, and that our Sun could be the same as the many stars littering the night sky. For these outlandish ideas he was found guilty of heresy and burned at the stake in 1600.

Bruno wasn't the first to question Earth's place in the grand scheme of things. Way back around 350 BC, the Greek philosopher Aristotle pondered whether our planet was part of some larger entity. But while he realised that the Sun and Moon are orbiting spheres, after a good bit of musing, he settled on the idea that Earth lay at the centre of the Solar System.

It wasn't until Nicolaus Copernicus (1473–1543) came along that Earth was knocked off its pedestal. Copernicus was a lawyer, but more crucially also a mathematician who enjoyed stargazing as an amateur astronomer. He worked out that if the stars were orbiting Earth every 24 hours, the more distant ones would have to be belting it through the cosmos at impossible speeds. So in his 1543 book *De revolutionibus orbium coelestium*, 'On the revolutions of the heavenly spheres', he proposed the controversial idea that the Sun lay at the centre of our Solar System – the so-called 'heliocentric model'.

The story goes that Copernicus was handed the book on his deathbed – he apparently awoke from a coma, took one look at it, then promptly expired. This was possibly just as well for him, for, as Bruno's fate shows, questioning heavenly design wasn't a good idea. However, Copernicus avoided the stake, possibly because he'd played the Church cleverly, by dedicating his book to the Pope.

The Final Nail

Now, it's all very well having an idea like Copernicus's, but to nail home the science, you need to collect the evidence to support it. Enter Galileo Galilei.

Galileo was an Italian mathematician, physicist and astronomer. In 1609, he heard of a new invention by a Flemish spectacle-maker. Galileo set about improving Hans Lippershey's 'spyglass' by grinding better lenses (see page 89). When Galileo turned this telescope skywards, a new world was revealed. He saw craters on the Moon and the rings of Saturn. But, crucially, he also spotted some odd shapes around Jupiter.

"When Galileo saw Jupiter had moons orbiting it, it confirmed the picture that Copernicus and others before him had dared to propose: that we human beings are not at the centre of the Universe," says Al-Khalili.

Galileo's evidence and support of Copernicanism was his eventual downfall – he ended his days under house arrest. But his book *Dialogue Concerning the Two Chief World Systems*, published in 1632, which looked at the arguments for different models of the Solar System, was smuggled out of Italy. This was the start of a new era in science, as, within a few decades, most educated Europeans came to believe that the Earth orbited the Sun.

THE VAST UNIVERSE

How a bitter rivalry revealed the
Milky Way to be just one of many galaxies

Nominated by **Sir Patrick Moore**, astronomer, author and
presenter of the long-running BBC show *The Sky at Night*

One night, when he was a young boy, Charles Messier caught sight of
a six-tailed comet in the dark sky. Three years later he watched a solar
eclipse from his hometown of Badonviller in France. From that moment
on, Messier was hooked – and he became a keen amateur astronomer. It
wasn't long before he'd got his own telescope, and was working under
the French Navy astronomer Joseph Nicolas Delisle, recording every
sight in the night sky to form a comprehensive catalogue. Published
in 1771, this documented star clusters and clouds of cosmic gas called
nebulae, which together became known as 'Messier objects'.

"There were over a hundred entries, but the objects were not at all alike.
The clusters, such as the Pleiades, were obvious enough. But the nebulae
were of two distinct types. Some, such as M42 in Orion's Sword, looked as
though they were gaseous, while others were starry," says Moore.

Astronomers were starting to suspect that the world above their
heads was more complex than previously thought. Then in 1845,

while peering through his 72-inch reflector telescope at his castle in Birr, Ireland, the Earl of Rosse spotted a strange object that looked like a modern-day Catherine wheel. It was, in fact, the first sighting of what we know today as a spiral galaxy.

The question was: how far away were these strange objects? At the start of the twentieth century, astronomers believed that the Universe was finite, and these nebulae and star clusters were no more than minor features in the Milky Way, the one and only galaxy. But whispers had started to turn into discussions that these nebulae were in fact other independent galaxies – or 'island universes', as they were nicknamed.

But 80 years passed before the discussion really heated up into a full-blown row between two stubborn astronomers.

Cosmic War

Edwin Hubble and Adriaan van Maanen disliked each other immensely. The problem was that they worked at the same observatory at Mount Wilson in California in the 1920s, so they kept bumping into each other. While the Dutch astronomer van Maanen believed the nebulae were local and existed in the Milky Way, Hubble thought they were located outside it.

"Van Maanen claimed that some of the stars in the arms of the spirals showed definite independent movement, which meant that they could be no more than a few tens of light-years away," says Moore. "Hubble had other ideas."

After a long time spent scanning the Andromeda 'spiral', Hubble at last found what he'd been searching for – Cepheids. These are a particular type of variable star whose brightness fluctuates over time. Studying their changes in brightness with a large telescope allowed Hubble to use them as a distance ruler.

"It had been shown that the real luminosity of a Cepheid is linked with its period [the time between two peaks in the star's brightness]. The longer the period, the more powerful the star," says Moore. "This means that once we know the period, we can find out its luminosity – and also its distance."

Having worked out the Cepheid distances using his super-strength 100-inch telescope, Hubble calculated that the 'spiral' was 8,000 light-years away – too far to be inside the Milky Way. We now know this is an underestimate of the distance to this 'spiral', known today as the Andromeda Galaxy. But this discovery forced astronomers to accept that the Universe was, in fact, millions of times bigger than previously thought.

So, how did van Maanen get it so wrong? He used a technique that compared objects near the edge of the nebulae. What he didn't account for were optical effects, which made the stars appear slightly closer than they really were.

"Van Maanen's photographic measurements were wrong, but completely honest mistakes," says Moore. "But, like other astronomers, he had to accept that our galaxy is one of many. The Universe is grander than had been generally believed, and our Earth is an insignificant speck. This happened less than a hundred years ago, and it marked a complete revolution in our outlook."

After this momentous discovery, Hubble used the results of American astronomer Vesto Slipher to deduce that these galaxies were, in fact, receding from us and from each other – the essence of Hubble's law. This hinted at the controversial idea that the Universe must be expanding and paved the way for the Big Bang theory.

THE BIG BANG THEORY

How a priest, pigeons and Planck proved the biggest explosion of all time

Nominated by **Nigel Henbest**, author of books
such as *The History of Astronomy*, and co-founder
of science documentary production company Pioneer

"The 'big bang' hypothesis is unpalatable, an irrational process that cannot be described in scientific terms." It was 1949 and the astronomer Fred Hoyle was sat in a BBC radio studio, arguing against the new-fangled theory that the Universe had formed billions of years ago from an event like no other – a vast explosion that left ripples of evidence, visible even today.

At the time, little did Hoyle know how ubiquitous his off-the-cuff nickname for the explosion would become. The Big Bang is integral to modern-day physics.

A brilliant astronomer and mathematician, Hoyle worked on numerous ground-breaking theories, from the panspermia hypothesis (that meteorites seeded life on Earth) to the theory of stellar nucleosynthesis (how the nuclei of elements form in stars). For a man who contributed so much to science, Hoyle, incredibly, never received a Nobel Prize. This may have been because he fell out with colleagues at Cambridge University and also with the Nobel Committee itself, but the more likely reason is that he became most famous for backing the wrong horse in what was to become in physics the greatest race of all time.

Throughout his life Hoyle supported the Steady State model of the Universe, in which the Universe has no beginning or end, with matter created continuously, like a steadily dripping tap. This contradicts the now well-accepted Big Bang theory, which states that in the first breath of the Universe's life (a mere hundred billionths of a yoctosecond – that's 10^{-31} or 0.0000000000000000000000000000001 seconds), it grew from an incomprehensibly small size, billions of times smaller than the tiny subatomic particle known as a proton, to about the size of a football pitch. In the next hundred millionths of a yoctosecond, the Universe formed a soup-like consistency made up of fundamental particles and antiparticles, before starting to cool from intensely hot temperatures in the trillions of trillions of degrees Fahrenheit, as the forces of nature, such as gravity, began to separate out. After 1 microsecond, the Universe had cooled down enough for protons and neutrons to form and then combine in the first few minutes into the atomic nuclei of helium and hydrogen.

For the next 300,000 years or so, still at a fairly toasty 100 million degrees Fahrenheit, the Universe appeared 'foggy', as particles interacted with protons. Then finally, when it had grown to a whopping 100 million light-years in diameter, the first atoms developed, as protons 'captured' electrons and dispersed through the cosmos as radiation.

Hubble's Cheat Sheet

It was a Belgian priest, of all people, who first suggested the idea that the Universe might have been created from one gigantic explosion. Working off the back of none other than Einstein's famous equation $E = mc^2$, Georges Lemaître proposed this idea to the great man in 1927. Einstein was sceptical, believing in a static cosmos.

But Lemaître's idea fitted in nicely with results that an American astronomer by the name of Edwin Hubble is credited with finding. Hubble's law states that the velocity of a galaxy is proportional to its distance from Earth. Hubble did measure the distance to galaxies, which hadn't been done before. But what he didn't do was work out the velocities.

"I'd like to set the record straight on this one," says Henbest. "Everyone says that Hubble discovered the Universe is expanding, but he actually took the results from a guy called Slipher."

In the early 1900s, American astronomer Vesto Slipher was investigating how spiral nebulae rotate. As he studied these amazing formations at Lowell Observatory in Arizona, it started to dawn on him that these nebulae were in fact rushing away from Earth. But it was Hubble who put two and two together, and using Slipher's results published a paper in 1929 that led to him being incorrectly credited with discovering cosmological 'redshift', in which light from a distant object is shifted more towards the red part of the spectrum the further it is from Earth. This redshift is due to the expansion of the Universe.

A good analogy for redshift is an ambulance siren. The closer an approaching ambulance comes towards you, the higher pitched the siren sounds. After it's passed by, the pitch sounds lower. Likewise, as a moving star or galaxy draws nearer, its light is shifted to the blue end of the spectrum; while the faster it goes away, the more its light is shifted to the red end.

Whether Slipher would have had a space telescope named after him is one thing, but you have to feel sorry for the man that 'Slipher's law' never came to be. Still, discovering radiation left over from the Big Bang in the form of cosmological redshift was momentous. It paved the way for confirmation that the Big Bang actually took place.

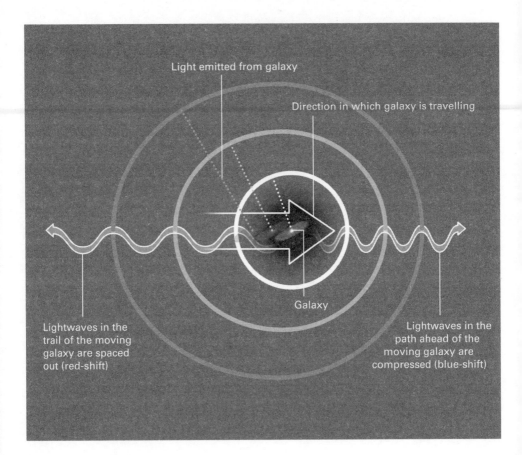

Light emitted from galaxy

Direction in which galaxy is travelling

Galaxy

Lightwaves in the trail of the moving galaxy are spaced out (red-shift)

Lightwaves in the path ahead of the moving galaxy are compressed (blue-shift)

Proof by Pigeon

Horrific as it is, war often leads to breakthrough innovations. The ancestor of radio astronomy was radar used during World War II. After the war, one of the first radio telescopes in the world was built by physicist Martin Ryle at a Cambridge facility. As he peered deeper and deeper into the cosmos, he realised that galaxies were bunched closer and closer together. There could only be one reason for this: that the Universe was smaller in the past. So, could it really be that if one looked back to the dawn of time, the Universe started as an extremely minuscule dot?

In 1964, the discovery by two American physicists added weight to this idea. Arno Penzias and Robert (Bob) Woodrow Wilson were scanning the skies with a radio telescope at their Bell Labs in Holmdel, New Jersey. Rather annoyingly, there was a persistent hiss from the telescope.

"At first Penzias and Wilson thought it was due to pigeon droppings in their telescope antenna. So they shooed the pigeons away," says Henbest. "The homing pigeons came back again and again and again. Eventually they were removed for good." But the hissing didn't stop. In fact, Penzias and Wilson found that this noise seemed to be coming from all over the sky.

At nearby Princeton University was Robert Dicke, who had been saying for some time that if there had been a Big Bang, there would be radiation left over from it. He'd been planning to build a telescope to look for this afterglow of creation. Fortunately, he got wind of Penzias and Wilson's results and got in touch, saving them all a lot of head-scratching – and earning the pair a Noble Prize for Physics in 1978.

Lumpy Universe

The final nail in the coffin for the Steady State theory was when the Cosmic Background Explorer (COBE) took to the skies in 1989. The satellite's mission was to peer into the past and capture the baby Universe. Three years later it caught a snapshot that would send a dagger into the heart of the Steady State theory.

For the Big Bang to have taken place, evidence of lumps and bumps in the fog of the early Universe should have existed. Previous probes into the 'microwave background radiation' which pervades all of space, weren't sensitive enough to pin down these subtle signs. But COBE went a step further than other space satellites. It probed deep space like an incredibly sensitive thermometer and discovered the oldest known objects in the Universe, so-called 'wrinkles' in time, that were the ancestors of today's galaxies. These could only be attributed to an enormous explosion – a Big Bang.

With the foundations laid, the theory grew with COBE's successor, the Wilkinson Microwave Anisotropy Probe (WMAP). "For me, the Big Bang is not a theory; it's a fact," says Henbest. "Observations first by WMAP and now by the Planck satellite prove it. Nothing else could be responsible for this radiation other than the afterglow of the Big Bang."

There's other evidence too. According to the theory, the gases created in the Big Bang should be made of 75 per cent hydrogen and 25 per cent helium. "That's exactly what the Universe is made of!" says Henbest. "And three different ways of dating the beginning of the Universe all agree. If you measure how long galaxies have been flying apart, by 'winding the tape backwards', you get a figure of about 10–15 billion years. The microwave background gives virtually the same result of 13.7 billion years. Plus, the age of the oldest stars work out at about 12–13 billion years. It can't be coincidence that all these ages are almost the same."

The Future

With the Big Bang now an established fact, one can only wonder what existed before. Have there been other Big Bangs, with universes expanding then contracting in a series of 'Big Crunches' over the aeons of time? And what of the future?

Astronomers now know that the Universe is not just expanding but, in fact, accelerating. The fault lies with an entity which as yet remains unexplained: dark energy. Alarming as an accelerating Universe sounds, we've still got many, many years – 1 followed by 100 zeros – before the Universe breathes its last breath.

DARK MATTER

Solving the Universe's biggest mystery

Nominated by **Lord Martin Rees**, Master of Trinity College, Cambridge, Astronomer Royal and former president of the Royal Society

Of all the matter in the Universe we can't 'see' over 80 per cent of it: we can only detect it through its gravitational effect. Astronomers have discovered that galaxies, and even entire clusters of galaxies, would fly apart unless they were held together by the gravitational pull of between five and ten times more material than we actually see.

"If we could solve the problem of dark matter – and I'm optimistic that we will within the next decade – we would know what our Universe is mostly made of; and we would discover, as a bonus, something quite new about the microworld of particles."
Martin Rees

Over the years, numerous different observations have come to this conclusion. One such is the evidence from disc galaxies. As the name suggests, these are galaxies that are disc-shaped, like our own. Stars and gas circle around the central hub of such galaxies, at such a speed that centrifugal forces balance the gravitational pull towards the centre. This is analogous – though on a massively larger scale – to the way the Sun's gravitational pull holds the planets in their orbits.

Radio astronomers can detect clouds of cold hydrogen, orbiting far out beyond the limit of the visible disc of stars. If these outlying clouds were feeling the gravity of the stars and gas in the galaxy, then the further out they were, the slower they would be moving, for the same reason that Pluto is moving more slowly around the Sun than is the Earth.

But that's not what has been found. Clouds at different distances from the galaxy all orbit at more or less the same speed. If, in our own Solar System, Pluto were moving as fast as the Earth, it would mean that there must be some shell of material outside the Earth's orbit but within Pluto's. Likewise, the high speed of these outlying clouds tells us that there's more to galaxies than meets the eye.

The entire luminous galaxy of stars and glowing gas must be embedded in a dark halo, several times heavier and more extensive. But what is this mysterious substance?

Gravitational Lensing

In the 1930s, the Swiss-American astronomer Fritz Zwicky argued that the galaxies in clusters would disperse unless they were restrained by the gravitational pull of some unknown matter. He suggested that gravitational lensing – the bending by gravity of light rays from objects behind it – could reveal the presence of this strange 'dark matter'. Many decades later, this technique has borne fruit.

There's no reason to question the idea that most of the stuff in the Universe may be dark – why should everything in the sky shine, any more than everything on Earth does? But the big question is, what could this dark matter be?

It's embarrassing to cosmologists that most of the Universe is unaccounted for. Earlier candidates for the missing matter included very faint stars or the dead remnants of stars. A whole host of arguments strongly suggest that dark matter isn't made up of ordinary atoms. There is still a lively debate about exactly what these mysterious and elusive entities could be, and there are no firm candidates. Physicists have theorised about many types of particle that could have been created in the ultra-hot initial instants after the Big Bang and which could have survived until the present.

Thousands of these particles could be hitting us every second, but they almost all pass straight through us. Sometimes one of them collides with an atom, however, and sensitive experiments might detect the consequent recoil when this happens within, say, a lump of silicon. Several groups around the world have taken up this challenge, but it needs incredibly delicate equipment deployed deep underground to reduce the background signal from things like cosmic rays, which are charged particles from space.

Dark matter is a key problem challenging astronomers today and it also ranks high as a physics problem. Astronomers, like Rees, believe they'll discover its true nature sometime within the next ten years.

EARTHRISE

How a photo changed our perception of our place in the Solar System

Nominated by **Dan Heaf**,
Digital Director of BBC Worldwide

When cosmonaut Yuri Gagarin rocketed into space on 12 April 1961, a new era of space exploration began. As NASA joined the space race with the *Apollo* missions, the world became entranced by the astronauts' tales of their cosmic adventures. While many of the audio recordings are poignant, without the images they shot in space few on Earth could have envisioned the world beyond our world.

"While the whole *Apollo* lunar project signified the achievement of technocratic goals," says Heaf, "I'd argue that the project's most enduring legacy is the collection of images taken from space during the *Apollo* missions."

The first of these photographs was captured on Christmas Eve, 1968, by William Anders on the *Apollo 8* mission – the first to escape our planet's orbit and see the far side of the Moon. 'Earthrise' showed the whole planet as it would appear from space, with the lunar horizon in the foreground.

"It reformed the way we thought about our planet as a whole, largely replacing the cartographers globe with its delineation of lands and seas on a gracticule of latitude and longitude," says Heaf. "It was also an image in the public domain, which saw it spread ubiquitously, almost viral before the web, in advertising and publicity."

Indeed, the collection of photos from all the *Apollo* missions are incredibly powerful and iconic images.

EXOPLANETS

Hunting for strange new worlds

Nominated by **Dr Chris Lintott**, astrophysicist
and presenter of BBC's *The Sky at Night*

Peering along the spiral arm of our galaxy, space telescope Kepler is on the hunt for new planets outside our Solar System – the so-called 'exoplanets'. Launched in 2009, it has already discovered hundreds of distant planets, some of which are truly bizarre.

From a distance, the exoplanet Kepler-7b seems to be swollen, with a density like that of a styrofoam cup, while CoRoT-7b orbits so close to its star that its surface is scarred by oceans of magma. Another exoplanet, gas giant WASP 12b, which is 1200 light-years from Earth, has a strange atmosphere. It differs from most

"The idea that our Solar System might not be typical really shook up my world. Until just a few years ago, we only had one solar system to study – our own – and yet out there there are a myriad of different systems. We shouldn't have been surprised, and yet the thought that there's this incredible diversity out there still sends a shiver down my spine"
Chris Lintott

planets we know by having an atmosphere dominated by carbon rather than oxygen, and so could have a tar-soaked surface and a graphite or even a diamond core.

Advances in telescope technology have enabled astronomers to catch glimpses of these weird worlds. To date, one of the most successful methods for locating them is by means of the 'wobble' method, used by ground-based telescopes. As a planet orbits a sun, its gravity 'pulls' the star back and forth. It's like what happens to an athlete in a hammer-throwing competition: as they whirl around, they get pulled off-centre by the weight of the circling ball. A planet is much smaller than a star, so the wobble is very subtle, but it can be picked up using the Doppler shift. The light gets blue-shifted if the star is pulled towards you, and red-shifted as it goes away again, just as an overtaking ambulance siren seems to rise and fall in pitch (see page 103).

Space telescopes such as Kepler use the 'transit' method for locating exoplanets; this works on the same principle as a solar eclipse. An exoplanet will cause the brightness of its parent star to dip ever so slightly as it passes in front of it. Provided that the telescope is in line with this 'eclipse', the exoplanet's presence is revealed.

Exotic Worlds

Using these techniques, astronomers are hoping Kepler and other telescopes will find even more strange new worlds, such as a cannonball-like planet or a waterworld. Depending on the initial ingredients in the mix, all sorts of interesting planets could develop.

Earth consists of a molten iron core encased in a mantle and crust of silicate rocks. Mercury is much denser, though, because it no longer has a silicate layer; this was probably blasted off by some huge impact. This could happen to a much larger exoplanet, leaving behind a cannonball-

LONELY PLANETS

Other exotic worlds that could harbour life are so-called 'rogue planets' that wander aimlessly through the cosmos. Gravitational tussels with passing siblings can cause planets to be 'flung' out of orbit and away from their parent star. But it appears that not all rogue planets are catapulted from their solar system.

Planets typically develop when gas and dust swirling around a young star smash together, accreting into bodies known as 'planetesimals'. As these collide, they form larger planets and eventually settle into orbits around the sun. It seems, though, that accretion isn't the only way to make a planet.

Isolated planets with no parent stars have been spotted in the Sigma Orionis star cluster. Intriguingly, astronomers discovered jets of gas streaming from their polar regions – a phenomenon normally seen when new stars form. So it appears that these exoplanets may also condense from an isolated gas cloud, just like in stellar formation, meaning they have no need for a parent star.

While rogue planets probably don't have sufficient heat to sustain life, it's possible that ejected exoplanets could maintain geothermal heat underground for long enough for life to develop or be seeded on another planet if a collision occurs.

like planet with iron-rich oceans and atmosphere. A waterworld might develop when ice and dust orbiting far from a star accrete to form an icy planet, which then thaws into water as it migrates closer to the hot sun.

From waterworlds to cannonball and tar-covered planets, there could be some exotic new worlds out there waiting to be discovered – some potentially with life on. It seems that fact could actually become stranger than fiction.

EXOWATER

One of the key ingredients for extraterrestrial life

Nominated by **Dr Lewis Dartnell**, astrobiologist, Centre for Planetary Sciences, University College London

It's a simple molecule, made up of just two elements – hydrogen and oxygen. It covers 71 per cent of our planet's surface and without it life as we know it cannot exist. The molecule in question is, of course, water.

In September 2009, NASA scientists confirmed that they'd found water ice in the shadowy craters at the Moon's south pole. The following year a radar instrument on an Indian space probe discovered in craters at the north pole a whopping 600 million tonnes of the stuff – that's the weight of about 110 million elephants.

However, while water on the Moon would be useful for a lunar base or human colony, our neighbour would never be able to support extraterrestrial life.

Water isn't the only key ingredient for life. You also need organic molecules – life's building blocks – and an energy source. Most life on Earth is solar powered, either directly by the Sun, as in the case of plants, or indirectly by eating those plants. There are exceptions, though: bacteria that live deep in the Earth's crust or in hydrothermal vents on the seabed. "If the Sun were to burn out tomorrow, these lifeforms would be quite happy subsisting off chemical energy pumping out of the Earth's crust – until the oceans freeze up," says Dartnell.

Next Stop: Mars

"Our next door neighbour is one of the best chances we've got for a habitable, or once inhabited, world in our Solar System," says Dartnell. "In many ways it's very Earth-like, or at least was in an earlier time."

Aeons ago, Mars was much warmer and wetter. Probes sent to Mars have beamed back a whole stack of different signs of liquid water, such as old river canyons and scars left by lakes. "With organic molecules and ample energy from the Sun, Mars essentially ticks all of the boxes of habitability and provides basic life support," says Dartnell.

But between 3.5 and 4 billion years ago tragedy hit. Astronomers aren't quite sure how it happened, but the planet lost its protective blanket of carbon dioxide that kept it warm. The Red Planet's environment collapsed, leaving it parched and excruciatingly cold at −100°C. Life literally turned to dust. Or did it?

There's a very small possibility that the 'seeds' of life remain dormant, awaiting just a few drops of water. Mars rovers that have sifted the dusty surface have never uncovered any signs of life, or even 'biosignatures' of life that once was. "Maybe life on Mars is still alive and kicking, but exists a couple of kilometres down, where it's still a bit warm from the internal heat of the planet," says Dartnell. "Just maybe there's a habitable niche on Mars today."

If not Mars, then maybe elsewhere in our Solar System. "Scientists are now wondering whether you could have life from other solvents, such as methane, which would get the biochemistry running, just like water does," says Dartnell. "For example, liquid methane exists on Titan, one of Saturn's moons."

Life on Exoplanets

If our Solar System doesn't yield any signs of life, then maybe it'll be found on one of the hundreds of planets that have been discovered elsewhere in the cosmos. "We now know that exoplanets are common as muck in our galaxy," says Dartnell. "Perhaps another world out there provides the perfect warm, wet conditions for life."

Apart from a solvent, energy source and organic molecules, other characteristics of a planet dictate its ability to support life. Size matters. If a planet or moon is too large, it may cling on to an atmosphere that's too thick. Or it might be too big to generate a magnetic field (see page 157), or perform processes like plate tectonics. "Plate tectonics and volcanism are a good way of regulating the atmosphere and climate," says Dartnell, "recycling the elements, like carbon, nitrogen and oxygen, through the biosphere."

If the exoplanet is near enough to a space-based telescope such as Kepler (see page 113), it can check it out for signs of life by analysing its atmosphere. If the gas concentrations in the atmosphere change over time, life could be there.

Astronomers have spotted water and the organic compound methane in the atmosphere of alien worlds, but these are on huge gas giants like Jupiter, which are not ideal to support life. "The holy grail is to find an Earth-like world with gases such as oxygen, carbon dioxide or methane in the atmosphere. Such small Earth-like exoplanets give off weak signals, so are difficult to see," says Dartnell. "Kepler made a substantial breakthrough in early 2011. It spotted 54 possible planets in the habitable zone of their star, five of which were 'SuperEarths'."

These 'SuperEarths' are planets that are thought to be several times larger than our Earth. The 'habital zone', also known as the 'Goldilock's Zone', describes the area around an exoplanet's parent star that's not too close or too far from it, so life doesn't freeze or boil. It also helps to have a satellite like our lunar brother to act as a stabiliser, so the planet doesn't swing wildly back and forth, and the climate stays stable for long periods of time. While 'simple' bacteria living deep on the ocean floor wouldn't care if the Moon didn't exist, 'complex' life like us needs long-term stability to evolve.

Exceptions to this rule may apply, though. Recent research at the University of Chicago suggests that planets ambling aimlessly through the cosmos could hold water, and hence life, on their surfaces. Gravitational battles with passing siblings can 'hurl' a planet out of orbit and away from its star. An Earth-sized exoplanet's internal heat would be enough to keep water liquid under ice sheets for at least a billion years. It's possible that these 'wandering' planets are stepping stones to seeding life elsewhere in the cosmos. But of this is still speculation.

Conan the Bacterium

While astronomers wait for more results from Kepler and other telescopes, Dartnell and fellow astrobiologists are pushing life on Earth to the extreme. After extracting microorganisms from acidic volcanic pools or the frozen deserts of Antarctica, they expose these extremophiles to a battery of tests to find out just how hardy they are. *Deinococcus radiodurans*, aptly-named 'Conan the Bacterium', can withstand severe desiccation and radiation exposure thousands of times higher than a human cell can survive. If Conan can endure this, it can probably put up with anything space has to throw at it.

In the near future, a small capsule will be strapped to a Russian probe on a mission to the Martian moon Phobos. Inside the 100 g capsule will be a bunch of extremophiles. The aim of this Living Interplanetary Flight Experiment (LIFE) is to see how these microrganisms will fare on their three-year round trip. The idea is that, if they can survive the extremes of space, they could survive a more homely environment of an Earth-like exoplanet.

The World of Physics

THE ATOM

The discovery of
the particles of life

Nominated by **Professor Paul Davies**, physicist, author,
broadcaster and director of BEYOND: the Center for
Fundamental Concepts in Science

The simplest of experiences can hatch eureka moments. Legend has
it that despite all his inherited wealth and global travels, the ancient
Greek philosopher Democritus hit upon one of the most fundamental
of ideas in physics while sitting in the comfort of his own home.

Reputedly, it was the smell of bread that instigated his groundbreaking
theory. As a servant walked upstairs with a fresh loaf, Democritus had a
brainwave – he deduced that for the smell to reach his nostrils, there must
be some minute particles of bread escaping from the loaf and wafting
through the air. If that was the case, then he surmised that a lump of
cheese could be cut in half, then half again, and so on. But eventually
there would be a minute piece of cheese that could not be cut in half, not
because there was no knife sharp enough, but because the final particle
was not something that could be sliced. He called this an atom, which
means 'uncuttable' in Greek.

Democritus believed that the Universe was made up of these atoms that flew around chaotically, billions upon billions of them bumping into one another and amalgamating to form round planets. The ideas of atoms and spherical planets were very advanced for 400 BC, and, as with any theory that's ahead of its time, if you can't prove it, it often gets forgotten – particularly if someone else shouts louder about a different idea.

"Of all the scientific knowledge we have about the world, the existence of atoms is the most fundamental and pervasive. It led inexorably to quantum mechanics which has decidedly changed the world through a myriad of technologies, such as the transistor and the laser. It led to molecular biology and the whole biotech revolution. It explained chemistry and is the basis for the industry that most impacts our lives on a daily basis. It led to nuclear power. And above all, it showed that human beings could concoct a sweeping theory of the physical world based on something that cannot be seen, and the existence of which took almost two and a half millennia to confirm."

Paul Davies

As Democritus passed away at the grand old age of 90, the ideas of a young Greek philosopher by the name of Aristotle were starting to be adopted. Aristotle claimed that all matter was made up of five elements: earth, fire, air, water and an aether (the upper sky). As a philosopher, he believed that all life was ordered and controlled by forces such as love and conflict, which completely contradicted Democritus's chaotic world. It would be another 1,800 years before the concept of matter being made up of minute particles would rear its head again – through the work of a mystery scientist.

Alchemist Anonymous

In the thirteenth century, an anonymous alchemist suggested that all bodies were made up of an inner and outer layer of minute particles, which he termed 'corpuscles'. Under the pen name Geber, the secretive scientist described how metals and compounds such as mercury and sulphur could be broken down into their constituent corpuscles.

This idea rumbled around the laboratories of science for a few centuries, but it was the groundbreaking work of Anglo-Irish scientist Robert Boyle, and then English scientist John Dalton, that really paved the way for the discovery of atomic theory.

In his aptly-named book *The Sceptical Chymist*, published in 1661, Robert Boyle suggested that all matter is made up of tiny particles joined together in different ways, and that materials could be made of more than one element – a compound. Over a century later, John Dalton elaborated on this idea, realising that different elements are made up of different atoms, and when these combine in specific proportions they form compounds. Dalton was the first to define exactly what an element was, as well as create symbols for each of the elements, which Dmitri Mendeleev later embellished on with his creation of the Periodic Table (see p185).

But, even as late as 1900, the physical reality of atoms was still rejected by influential scientists, most notably the Austrian physicist Ernst Mach and the German chemist Wilhelm Ostwald. Their refusal to countenance the existence of atoms was based largely on a Positivist agenda, in which the lack of direct evidence for atoms – and the supposed impossibility of obtaining any – in essence implied their non-reality. This view led them to mount a sustained and vociferous campaign against the views of the Austrian physicist Ludwig Boltzmann, who had shown that the presumption of the physical reality of atoms led to natural explanations for the bulk properties of matter.

Boltzmann and his work was successfully marginalised for many years, and by the time of his suicide in 1906 he was regarded as a scientific 'dinosaur'. Ironically, barely a year before his death, a paper appeared that ultimately established the reality of atoms. It was an analysis of the phenomenon of so-called Brownian motion, the random movement of particles in a suspension, which was shown to be explicable by the existence of atoms. The author of the paper was a young patent clerk named Albert Einstein. Within two years of Boltzmann's death, experimental studies of Brownian motion had compelled even Ostwald to accept the reality of atoms.

The first insights into the nature of atoms emerged in 1897, when British physicist Joseph John Thomson saw how so-called cathode rays passing through electric and magnetic fields bent towards a positively charged electric plate. It dawned on him that these must be negatively charged particles – and, more crucially, that the particles must be coming from within the actual cathode. He revived the name 'corpuscles'. Through his extensive series of experiments, Thomson had discovered electrons and so deduced that atoms are made up of even smaller particles.

The Plum Pudding Model

Reportedly, in his Cambridge laboratory, Thomson toasted the electron's discovery with the words: "To the useless electron." In fact, this 'plum pudding' model, so-nicknamed because the particles were thought to be like plums in a pudding, won Thomson the 1906 Nobel Prize for Physics and proved helpful in understanding key physical phenomena, such as electricity.

In the following years, a number of physicists and chemists worked together to identify more of the structure of atoms. Physicists Ernest Marsden and Hans Geiger tried firing all sorts of particles at various substances, measuring the deflection angle. Their results showed that the plum pudding model didn't quite ring true, as there was a central mass which deflected the rays at a greater angle than the external parts of an atom. In 1911, the physicists' mentor New Zealand-born physicist and chemist Ernest Rutherford suggested that the atom was made up of a dense nucleus orbited by electrons.

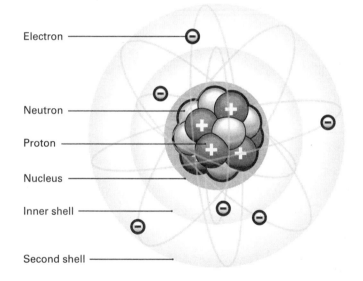

From this concept of electrons whizzing around a central mass, a bit like planets orbiting the Sun, another of Rutherford's students called Niels Bohr then devised a system for the structure of atoms. Bohr's model proposed that electrons orbit the nucleus in fixed 'shells'. Each element's chemical properties depend on the number of electrons in the outer shell. But, in 1918, Rutherford cracked a problem that had been puzzling them all for a while.

The way in which elements reacted didn't coincide with their masses. Rutherford solved the problem by working out that the negative electric charge of an electron was balanced by a positive charge in a particle called a proton. The remainder of the atom's mass was made up by 'neutrons' with a neutral charge, which, like protons, are found in the nucleus.

To give an idea of scale, the nucleus would be the size of a pea in an atom the size of the Albert Hall, with electrons the size of grains of sand whizzing around the stands. Maybe it is not surprising then that it took another 14 years for the neutron to be found and its existence confirmed, winning James Chadwick a Nobel Prize in 1935.

We now know that there are even smaller subatomic particles that make up matter, such as quarks and gluons. It's not actually possible to see these – instead, their behaviour and properties have to be gauged by studying the debris left over from particles that contain them.

Deep below the French-Swiss border the world's most powerful particle accelerator, the Large Hadron Collider, is revealing them by colliding 'beams' of particles accelerated inside a 27 km-long tunnel. By analysing their properties, scientists hope to understand the life story of matter in the Universe.

QUANTUM THEORY

How the subatomic world was revealed

Nominated by **Professor Michio Kaku**, theoretical physicist, author and presenter

"Anyone who is not shocked by quantum theory has not understood it." So said Niels Bohr in 1958. The Danish physicist wasn't alone in being baffled by it. One of the biggest challenges for twentieth-century scientists was to get to grips with the bizarre implications of the rules that govern all that occurs in the subatomic world.

Towards the end of the nineteenth century, physics was in a state of crisis. One mystery centred on finding laws governing the light and heat emitted from hot, glowing bodies such as the Sun. Such laws were potentially immensely useful: they would, for example, allow the Sun's temperature to be worked out simply by measuring the amount of solar energy that reaches the Earth.

"Quantum theory makes possible lasers, computers, the transistor, the MRI machine and all the wonders of the computer and information age. Without quantum theory much of modern science would not exist."

Michio Kaku

To make the search easier, during the mid-1800s physicists had invented an idealised object, which absorbed all the heat and light falling on it. Understanding the laws governing such an object – known, rather unimaginatively, as a black body – was expected to resolve key issues surrounding real-life glowing objects. But then a major problem reared its head. Two British physicists, Lord Rayleigh and James Jeans, independently found that the amount of energy emitted at high frequencies was infinite. So serious was the problem that it was called the 'ultraviolet catastrophe'. The situation was eventually resuced by Max Plank, a German physicist working at the University of Berlin.

Rayleigh and Jeans had supposed, in keeping with classical physics, that the black body absorbed and emitted energy in a continuous fashion. In 1900, Planck proposed a bold and revolutionary idea: that the absorption and emission processes were not continuous but occurred in discrete packets or bundles. These he called 'quanta'. The energy of each quantum depended only on the frequency of the radiation and a constant, the now famous Planck's constant. This limited the energy present at high frequencies and the ultraviolet catastrophe was averted. Also, the black-body spectrum calculated by Planck was in perfect agreement with experiments carried out. Planck's theory marked the birth of the quantum age.

Another phenomenon that troubled physicists in the early twentieth century was the so-called 'photoelectric effect'. It was observed that some metals such as zinc emitted electrons when exposed to light. The mystery was in the way the electrons were emitted. Low-frequency light, no matter how intense or energetic, would not dislodge electrons from the metal; yet high-frequency light of low intensity or energy would dislodge the electrons. In 1905, the scientific genius Albert Einstein used Planck's ideas to interpret these curious results. All radiation, he suggested, was quantised: light was composed of discrete packets of energy, or quanta, which behaved like particles. Einstein realised that the electrons could only be dislodged if the individual light quanta possessed sufficient energy to eject them and this in turn depended on the frequency of their oscillations. Later these particles of light were to be called 'photons' and play an important role in the physics of electromagnetic radiation. It was for this work, and not for relativity, that Einstein won the 1921 Nobel Prize in physics.

At around the same time, physicists were wrestling with another puzzle, this time concerning the behaviour of electrons in atoms. Classical physics predicted that the electrically-charged electrons orbiting around the nucleus should radiate energy, which should make them collapse into the nucleus in a fraction of a second. But the very fact that atoms –and we – exist proves this wasn't happening.

The young Danish physicist Neil Bohr had an idea – maybe electrons simply didn't obey classical physics, but could only orbit at certain fixed distances from the centre of the atom. This in turn meant that they could release only fixed 'packets' of energy as they jumped between orbits, released as photons of light. Although Bohr's ideas explained a number of curiosities about the atom, his model still lacked a really convincing foundation. The physicist Otto Frisch joked that Bohr's model required an 'atomic policeman' to ensure that electrons occupied only those orbits that were allowed.

The next development in quantum theory came from a French nobleman, Prince Louis de Broglie, who suggested that if light, which was normally considered a wave, could be treated as a stream of particles, then perhaps particles such as electrons could be treated as waves. In 1924, he proposed a formula which associated a wavelength to a particle in terms of its momentum, and with this formula he managed to explain Bohr's allowed orbits: when a particle moves in a circular orbit, the corresponding wave must link up with itself, like a snake eating its own tail – the circumference of the circle must contain a whole number of waves. Thus Bohr's forbidden orbits simply couldn't exist. The idea that subatomic particles could be treated as both particles and waves was a very important idea, and it was an Austrian physicist who made it yield results.

In 1925, Erwin Schrödinger devised a wave equation for an electron trapped in an atom that fitted de Broglie's formula relating wavelength and momentum. He then applied this equation to the simplest atom, that of hydrogen, with great success. Schrödinger's model of the atom was quite unlike Bohr's. There were no definite electronic orbits, just a fuzzy, pulsating cloud. The electronic charge was smeared out throughout this cloud instead of circling the nucleus in small, neat packets. The particular state an electron is in is described by its 'wave function'. When an electron makes a transition from a higher to a lower energy state, with the emission of a photon, there is no sudden jump, as with the Bohr model. During emission, the wavefunction consists of a mixture of the two wavefunctions corresponding to the initial and final states. This idea, known as the principle of superposition, means that the electron is effectively in two states at the same time! Once the photon is observed, this mixed-up wave function is said to 'collapse' to that of the final state.

Max Born introduced the concept of probability into quantum theory. Schrödinger's wavefunctions, he suggested, could be regarded as indicating the likelihood of an electron being at a particular place in the electron cloud at any given time.

A little before Schrödinger produced his wave mechanics, the German physicist Werner Heisenberg came up with an alternative description of the atom based on mathematical objects called 'matrices'. In his formulation, the fuzziness was expressed by his now fairly well-known uncertainty principle: if we know where an electron is, then we don't know what it's doing; and if we know what it's doing, we don't know where it is.

In 1927, the great English physicist Paul Dirac produced the definitive formulation of the general principles of quantum theory. His was not only a more general mathematical picture than either Schrödinger's or Heisenberg's, but it also showed that their approaches were equivalent.

The equations of quantum theory give us the most accurate results of any scientific theory. Yet there is something disquieting about how we are to interpret what is going on. Planck's constant is extremely small, so quantum effects do not generally feature in the world of everyday reality. How is the strange quantum world – where entities are not either waves or particles, but can be both at the same time, and where objects are not either here or there, but a bit here and a bit there – to be reconciled with the world of everyday experience.

The standard interpretation, devised by Bohr and called the 'Copenhagen interpretation', is that it is the process of observation or measurement that produces the definite result we obtain in the world of everyday reality. This interpretation bothered Schrödinger, who devised a dramatic thought experiment to highlight the nature of the problem (see box on page 136). The most bizarre interpretation of quantum measurement, however, is that proposed by the American physicist Hugh Everett III in 1957 and called the 'many-worlds' interpretation. This states that, in every situation where there are a number of possible outcomes, the world splits up into many parallel worlds, in each of which one of the possible outcomes actually occurs. In this interpretation, Schrödinger's cat has expired in one universe, but is alive and well in another. It is, of course, absurd to suggest that the cat's well-being depends on an observer. What Schrödinger was really asking was: when does the jump to 'definiteness' take place?

We can seek some solace in the words of the great American physicist Richard Feynman: "I think I can safely say that nobody understands quantum theory."

SCHRÖDINGER'S CAT

A cat is confined to a sealed box, along with a radioactive source and a vial containing poison. The radioactive source has a 50/50 chance of decaying in the next hour, emitting some sort of radiation in the process. If the emission takes place, the vial breaks, releasing the poison, which kills the cat. According to the Copenhagen interpretation of quantum theory, after an hour, the cat is in a superposition of the states 'alive' and 'dead'. When we look in the box, the wavefunction 'collapses' and we see the cat either alive or dead.

PARTICLE ACCELERATORS

Smashing particles in search of the very fabric of the Universe

Nominated by **Dr Steve Myers**,
Director of Accelerators and Technology at CERN

When the world's largest particle accelerator, the Large Hadron Collider (LHC), was switched on for the first time in September 2008, mavericks started predicting the end of the world was nigh. Apparently, we were all going to be sucked into one of the miniature black holes due to be created in the 27 km-long tunnel deep below the French–Swiss border at the LHC run by CERN. But we're still here.

"By accelerating particles to very high energies, particle accelerators and colliders allow scientists to recreate, in a controlled way, and study the conditions which existed a billionth of a second after the creation of the Universe. These studies will allow mankind to understand the fundamental constituents, forces and laws of nature."
Steve Myers

Despite a lengthy hiccup in 2008, when an electrical fault caused a 14-month shutdown, the LHC is now churning out useful data that confirm key features of current theories on how the cosmos came to be. The search is still on for elusive subatomic particles, explanations to various enigmas like dark matter (see page 108), and the answers to fundamental questions like why atoms have the properties they do.

The LHC is just one of many particle accelerators around the world that are trying to answer these questions. Physicists hope to glean the answers using brute force, by smashing particles together at energies close to the speed of light. Most particle accelerators are made up of circular underground tunnels through which subatomic particles are steered by extremely powerful magnets on the tunnel walls. By cranking up the magnetic strength, two opposing 'beams' of particles are accelerated towards one another. Making them collide head-on is no easy matter. It's like firing two needles at one another from 10 km apart.

It's not possible to see individual subatomic particles; instead, their behaviour and properties have to be gauged by studying the debris left over from the collision. By photographing the particles' paths as they pass through a magnetic field, physicists can identify them: positively charged particles swerve one way, negative ones the other, and their mass determines the degree to which their tracks are curved by the magnetic field. Physicists already know the characteristics of a whole host of particles, but it is hoped that the LHC will soon reveal elusive subatomic particles displaying new characteristics. They are on the look out for one particle in particular – the Higgs boson, aka the 'God particle'.

Hunting for the God Particle

Since the discovery of atoms, physicists had been baffled by how an atom could be heavy. It was easy to understand how an elephant, car or even a pen might have mass, as they're all made up of atoms. But where did an atom's mass come from? English theoretical physicist Peter Higgs worked out the answer while walking in the Scottish Highlands in 1964.

Higgs realised that particles seem massive because they're in fact battling their way through a force field – the 'Higgs field' – a bit like a celebrity trying to get through a crowd of photographers, who cram in ever-closer to get the best shot. The Higgs field is made up of subatomic particles called Higgs bosons. As a particle with no mass travels through space, it distorts the field as Higgs bosons cluster around it. This means that the Higgs boson gives all other particles their masses. Different particles attract different numbers of Higgs bosons, so that's why they have different masses.

While other particle accelerators have narrowed down the search for the Higgs boson, none of them has been powerful enough to find it. The LHC is on a different scale: its largest particle detector is as tall as a seven-storey building and its biggest magnet weighs as much as five jumbo jets and can store enough energy to melt 18 tonnes of gold. With stats like these, it is hoped that if any accelerator is going to find the Higgs boson then the LHC will.

But it's a tall order. Again, there's no chance of seeing the Higgs boson directly, as it decays almost instantly. It's also unclear at precisely what collision energies the Higgs boson should appear. With particle collisions occurring 600 million times every second, maybe it's already been produced at the LHC, but its existence was lost in the plethora of other data.

Physicists are confident that the LHC will eventually find the Higgs boson. If discovered, it would be the final piece in the jigsaw of the so-called 'Standard Model'. This model is the closest that theorists have come to the Theory of Everything, which describes all known phenomena, such as familiar particles like electrons, plus all but one of the fundamental forces that act on them – the exception to the rule being gravity.

If the LHC doesn't find the Higgs boson, theoretical physics would be plunged into crisis and our understanding of the particles making up all atoms in our Universe may need to be rewritten.

Two pipes travel through the 27km-long tunnel of the LHC. Each carries a beam of particles travelling in the opposite direction. These beams can be made to collide head-on where the detectors are located. Superconducting magnets are used to steer and focus the particles.

LHCb – Here the behaviour of particles called beauty quarks and anti-quarks are studied. The former behave like conventional matter, while the latter are a form of anti-matter – matter which has the same mass but the opposite electric charge. Both are incredibly unstable, but behave slightly differently when they disintegrate. The LHCb is designed to study these differences.

CMS – Like ATLAS, the Compact Muon Solenoid is a general-purpose machine, but has a completely different design and measures the mass and velocity of particles in a completely different way. This gives a crucial independent check of evidence emerging from ATLAS.

ALICE LHC – Studies the Higgs boson and supersymmetry. This detector studies the outcome of smashing together fragments of lead atoms until their protons and neutrons break apart. The collisions create a 'soup' of particles similar to that which existed just after the Big Bang.

ATLAS – At 7,000 tonnes and 25 metres high, ATLAS is the biggest particle detector ever built. It contains six different devices and a colossal magnet that together are expected to find evidence for the two principal targets of the ALICE LHC: the Higgs boson and supersymmetry.

The two final detectors are the **TOTEM** experiment. These measure the size of protons and how precise the collisions are in the LHC and the LHCf detector, which simulates cosmic rays.

ANTIMATTER

The science fiction idea with profound consequences

Nominated by **Tara Shears**, physicist and Royal Society University Research Fellowship holder

'Antimatter drives' in spacecrafts are pure science fiction. But the actual idea of antimatter is not. These strange particles, which are the mirror image of particles of matter, have been intriguing and baffling physicists for decades.

Before the 1920s, Schrödinger's wave equation described slow-moving quantum particles (see page 131), but it didn't explain particles travelling at the speed of light. Then, in 1927, a young physicist hit upon a solution.

The brilliant British physicist Paul Dirac hit upon a version of Schrödinger's wave equation which

"Antimatter is a bizarre and counter-intuitive idea to have in the first place, yet one of fundamental importance in explaining the evolution of the Universe."

Tara Shears

worked for both the framework of relativity and quantum theory in the case of the electron. But, as so often happens in life, the more knowledge you gain, the less you know – and this case proved to be no exception on that front.

Dirac's equation gave an unexpected extra solution that implied the existence of antimatter electrons. Normal electrons are negatively charged, while antimatter ones are positively charged. At first, these so-called 'positrons' were only predicted by the Dirac equation. But five years later American experimental physicist Carl Anderson actually found them, while tracking showers of particles caused by the cosmic rays which smash into the atmosphere from space.

We now know that every particle has an antiparticle. When they collide, they annihilate each other, and both their masses are converted into a photon of electromagnetic radiation.

Theory says that matter and antimatter should have been produced in equal amounts when the Universe formed. It seems, though, that just one second after the Big Bang, most antimatter disappeared, leaving behind only matter. This suggests that something is missing from our current theory, but we don't know what it is.

On the hunt for the answer are physicists working at CERN, who have been creating antimatter atoms for many years. The difficulty has been capturing the particles for long enough to analyse them. But,

DID YOU KNOW?

- A millionth of a gram of antimatter could have the potential to power a year-long trip to Mars.

- A year's worth of antiprotons would barely be enough to provide us with three seconds of electrical light.

in June 2011, researchers at the Large Hadron Collider succeeded in trapping antimatter atoms for 1,000 seconds – long enough to study in detail the properties of these mysterious particles.

The hope is that physicists will be able to discover more about the moments after the Big Bang, and determine whether our current theory is correct. As Shears says: "Antimatter is key to everything from explaining the Big Bang to being used to date rocks and locate tumours – something that Dirac would never have imagined."

GENERAL RELATIVITY

The journey to $E = mc^2$

Nominated by **Dr John Gribbin**, Visiting Fellow in Astronomy at the University of Sussex, and author of books including **The Reason Why**

The well-known tale that Isaac Newton discovered the concept of gravity by watching an apple fall from a tree, is almost definitely untrue. He likely dreamt up the story himself to ensure he gained credit for defining gravity, and to cement his name in the annals of science. It worked. But Newton did stand on the shoulders of a number of giants to create his theory of gravity.

The Italian polymath Galileo Galilee first conducted a series of experiments to determine the nature of how different objects fall. Reputedly, the experiments involved dropping two weights from the top of the Leaning Tower in his home town of Pisa. While the story about the tower is likely to be the result of centuries of embellishment, Galileo did work out that objects of different sizes and weights fall at the same speed.

Newton's big breakthrough was to realise that the same phenomenon occurred in space. He succeeded in calculating the orbits of the planets because of the law devised in 1609 by the brilliant mathematician and astronomer Johannes Kepler, who had worked out that all the planets move in elliptical orbits. Around 75 years later, by applying calculus (see page 314) to Kepler's first law of planetary motion, Newton successfully worked out the force of gravity needed to lock the planets in their orbits around the Sun.

"Newton made a vital contribution to science when he realised that the whole Universe is governed by the exact same law of gravity, whether it's an apple falling from a tree or the planets orbiting the Sun, and not something local to our own little planet," says Gribbin. "If the laws weren't the same everywhere, we wouldn't have a clue what was going on – and we probably wouldn't be here to wonder about it anyway, since it would be hard for life to evolve in a lawless universe."

Relativity

It was Galileo who first proposed in 1632 that the laws of physics are the same for all observers moving at constant speed – the essence of Relativity. For example, if someone is swinging a yoyo on land, someone else on a boat motoring by will see it making the same movements. Relativity is the reason that you can't tell if you're going forwards or backwards on a moving train if you've got your eyes closed.

Galileo's theory worked most of the time but, by the end of the nineteenth century, scientists regularly started to spot inconsistencies cropping up between the laws of mechanics and electromagnetism equations (see page163). A new solution was needed.

A young physicist, working in a patent office in Bern, Switzerland, came up with a revolutionary idea in 1905. Albert Einstein realised that the laws of mechanics broke down when describing moving objects close to the speed of light. It led him to suggest that the speed of light is actually constant for all observers and isn't dependent on how fast someone is travelling themselves. But there was one problem niggling at this theory of Special Relativity – it didn't fit with the gravitational force described by Newton's theory of gravity, which stated that nothing can exceed the speed of light. This bugged Einstein for almost a decade until, in 1915, he cracked the problem. In his General Theory of Relativity, he came up with the concept of curved space-time. He envisaged space to act a bit like a sheet of rubber, where a marble rolled across its surface would go in a straight line. But if a much larger object like a bowling ball was placed on the rubber sheet, its presence would make the marble's path curve. This is how gravity can be explained – massive objects affect the path of smaller objects. Einstein's General Theory of Relativity was published in 1916 and tied together space-time and gravity.

While forming his theories on relativity, Einstein also came up with his famous equation $E = mc^2$ which describes that an object's energy (E) is just equal to its mass (m), multiplied by the speed of light (c), squared. He'd had to add in a 'cosmological constant' to make the equation fit with the view at that time that the Universe was static. This fudge annoyed Einstein immensely, and he called it his "biggest mistake".

But Einstein should have had more faith in himself. We now know that the Universe is expanding – and the cosmological constant is necessary to describe this expansion.

THE LAWS OF MOTION

The backbone of
modern-day physics

Nominated by **Professor Frank Close**, professor of physics at
Oxford University and author of *Neutrino*

Isaac Newton didn't like to share the scientific stage. A portrait of
Robert Hooke suspiciously went 'missing' when the Royal Society
moved to new premises in 1710. Rumour has it that his arch-rival and
then president, Newton, ordered the painting to be destroyed. Without
an image of Hooke on the wall for future generations to see, his life's
work could have slipped into obscurity.

Fortunately, it didn't. Hooke is remembered not only as the chief
assistant to Christopher Wren, helping to rebuild London after the
Great Fire of 1666, but also for his enormous contributions to science,
from coining the term 'cell' to discovering the law of elasticity.

But Newton's own PR machine ensured Hooke is forgotten for being
the first to describe one of the laws of motion. In Hooke's 1674 book *An
Attempt to Prove the Motion of the Earth by Observations*, he wrote:
"All bodies whatsoever that are put into a direct and simple motion,
will continue to move forward in a straight line, till they are by some

THE LAWS OF MOTION

First Law: Every body remains in a state of rest or uniform motion unless it is acted upon by an external unbalanced force. This basically means that if an object like a ball is rolling along on a flat surface, it will keep rolling at the same speed unless something stops it, such as another object like a wall or friction with the surface it's rolling on.

Second Law: The force acting on a body is equal to its rate of change of momentum, which is the product of its mass and its acceleration. So, you need to use more force to accelerate a 10 tonne truck than a 10 g toy car.

Third Law: For every action there is an equal and opposite reaction. For a rocket to take off, hot gas is forced downwards, propelling the rocket upwards.

effectual powers deflected." This sounds an awful lot like Newton's First Law of Motion, which he didn't propose until over a decade later, when he published his *Principia* in 1687.

In fact, it was the Italian polymath Galileo Galilei who first hinted at this phenomenon: "A body moving on a level surface will continue in the same direction at a constant speed unless disturbed." He was essentially noting that if a ball rolls through treacle, for example, it comes to a stop almost straight away. If it rolls through water, it will keep going but slowly. If it rolls through air, however, you hardly notice that it slows down.

"From this, Galileo realised that as the stuff through which the ball travelled got thinner and thinner, the ball carried on for longer and longer," says Close. "From that he extrapolated the idea that if you had a vacuum, the ball would carry on without impediment."

This was a remarkable discovery for its day. "You'll remember thinking as a kid that objects just naturally stop of their own accord – you push something and eventually it comes to a halt," says Close. "So it's counterintuitive to discover that an object continues at the same velocity in a straight line unless a force acts on it."

On the Shoulders of Giants

Scientific breakthroughs often happen when scientists work from other scientists' results. Newton himself acknowledged this when he wrote to Hooke in 1676: "What Descartes did was a good step. You have added much several ways... If I have seen a little further it is by standing on the shoulders of Giants."

It's debatable as to whether this was a sarcastic remark meant to belittle Hooke, who suffered from a stoop, or whether he actually had the generosity to acknowledge the other man's greatness. Newton was regularly entangled in unpleasant disputes with other scientists. His cantankerous character could be blamed on an unhappy childhood, and being sent away from home to be educated.

Whether he was an unpleasant character or not, Newton's contributions to science are undisputed. He was the first to lay out a set of mathematical equations for the laws of motion. It's these laws that have moulded much of modern-day physics, informing Einstein's theories of general relativity and space-time, and leading to many technological advancements. As Close says: "Newton actually identified the laws of motion, enabling us to build machines and send spacecraft to the Moon."

ELECTRICITY

How a lightning bolt and a kite demonstrated the nature of electricity

Nominated by **Andrew Cohen**, head of the BBC Science Unit

A popular party trick at the start of the eighteenth century was to charge yourself up using a spark generator, then mingle with guests giving them electric shocks.

This static electricity was first noted way back around 600 BC by Greek philosopher Thales of Miletus. He recorded how rubbing fur on fossilised tree resin (amber) made the amber attract things such as feathers. Just as rubbing a balloon on a wolly jumper builds up static, so loosely bound materials, such as fur and feathers, have electrons which can develop an electrical attraction to the likes of amber, which has tightly bound electrons and easily develops a negative charge.

The word 'amber' in Greek is *elektron*, which led to the word 'electricity', after Queen Elizabeth I's physician William Gilbert used the term 'electricus' in his work *De Magnete*, published in 1600. Throughout the seventeenth century, many scientists played around with magnetism and electricity. Then, in the mid-1740s, the world's first capacitor was invented. This large glass jar, partially filled with water, would build up a charge from the friction mechanism inside. Linking a number of these jars together could create a massive charge.

The usually mild-mannered Benjamin Franklin was one of the scientists who couldn't resist the fun that could be had with a capacitor. He's known to have served his guests electrified champagne or turkey that had met its demise through an electric shock. This incredibly successful master of all trades was to play a significant role in the history of electrical discovery.

Born in Boston in 1706, Franklin's genius was noted at a young age. At the age of ten, his school funds ran out and he started working for his father, who owned a business making wax and candles. Bored by this, he then worked for his brother's printing company and, after a brief stint working as a printer in London, returned to the US to set up a successful publishing house in Philadelphia.

Next, politics beckoned, and Franklin became a key figure in the American Revolution; he's remembered as one of the Founding Fathers. He was also a prolific inventor, coming up with all sorts of ideas from bifocal lenses in glasses to swimming fins. Franklin's inventive mind was intrigued by electricity, and he started investigating the phenomenon seriously, not just in order to surprise his dinner guests.

In those days, lightning was seen as a sign of the wrath of God. Franklin set out to prove this wasn't the case but that it was in fact an electrical current. Ensuring that he was well insulated for fear of electrocution, he ventured out in the middle of a storm and hoisted into the air his makeshift kite: a silk handkerchief strapped to some sticks. At the end of the wet piece of kite string was a metal key.

As lightning struck the kite, an electrical charge was carried down the string and sparks flew off the key. Franklin had proved that lightning bolts are simply enormous electrical sparks. From this discovery, he invented the lightning rod that is attached to tall buildings, a design that has barely changed to this day.

Franklin's input into the world of electricity didn't end there. He went on to coin the terms 'charge', 'conductor', and also 'battery'.

THE BATTERY

How twitching frogs' legs led to the creation of the battery

Nominated by **Professor Daniel Dennett**, professor and co-director of the Center for Cognitive Studies at Tufts University

Big discoveries often happen by pure accident. One day in 1771, Italian physician Luigi Galvani was chopping up a dead frog in an attempt to prove that its testicles were in its legs. As he touched the frog's legs with his metal instruments, he noticed they twitched slightly. Galvani claimed this was due to an 'animal electricity' inside the frog.

Another Italian, physicist Alessandro Volta, had a different idea. He showed you could also make frogs' legs twitch by attaching two different metals to the frog's leg to form a circuit in which an electric current flows. The same process happens if you get a piece of foil stuck between your teeth and a metal filling. The two different types of metal produce a small electric current, which is why the foil feels uncomfortable in your mouth

Volta is credited with inventing the first battery in 1800. His 'voltaic pile' was so-called because it was made up of a pile of silver and zinc discs, each separated by a piece of cloth soaked in salty water. The chemical reactions in the pile generated a steady electric current equivalent in power to an AA battery.

Baghdad Battery

In 1938, a German archaeologist unearthed a clay jar near the city of Baghdad in Iraq. The jar, dated to around AD 200, contained an iron rod surrounded by a copper cylinder, and there were signs that it had been filled with some kind of acidic liquid, like vinegar. All sorts of applications have been suggested, from decorative electroplating to some kind of medical electrotherapy. If it had been used to generate an electric current, this 'battery' would pre-date Volta's pile by 1,600 years. Considering Galvani's unwitting discovery, it wouldn't be hard for someone to have discovered the same by mistakenly sticking a pair of copper and iron pins into, for example, a lemon. The jury is still out on this one.

Mass Production, Mass Impact

After Volta created his first battery in 1800, almost another century passed before batteries were mass-produced, going on sale in 1896. Since then, their impact on society has been vast – from TV remote controls to cars. And, as Daniel Dennett points out, batteries will continue to be crucial in the future: "Batteries are still no match for the mitochondrial ATP system the human body uses to power itself, but the explosion of science and technology that will allow us to colonise space depends on our ability to store and extract power easily, which we can do with batteries."

DIFFRACTION

Using light to elucidate the structure of matter

Nominated by **Professor Bill Jones**, head of the chemistry department at Cambridge University

Grimaldi crater on the Moon is named after the Jesuit priest Francesco Maria Grimaldi, not for his divine calling, but because he was a great mathematician and physicist, becoming a professor at the university in his home town of Bologna in Italy. He is credited with being the first person to accurately observe and record the diffraction of light, although some claim Leonardo da Vinci beat him to it. However, Grimaldi was one of the earliest physicists to describe the wave-like nature of light and it was he who coined the term 'diffraction' in 1665.

Diffraction occurs with all types of waves: sound, water, visible light, radio and X-rays, and so on. It's a

"Understanding the structure and properties of elements and compounds is vital for the development of new materials and technologies. Diffraction has applications from semiconductors to DNA."

Bill Jones

form of interference, where multiple waves interact with one another, causing them to bend and spread out when they encounter objects in their path. It's at its strongest when the objects are similar in size to the wavelength of the waves. It can be seen in the spreading of waves from the bow of a ship, to the colours on a CD or DVD caused by light interacting with the closely spaced pits on its surface.

Diffraction of X-rays has proved extraordinarily important, as the wavelength of X-rays is comparable to the spacing of atoms in crystals. As such, X-ray diffraction has been used to probe the atomic structure of everything from proteins to new materials – and even lethal viruses, such as HIV. Perhaps its greatest triumph came in 1953 when James Watson and Francis Crick used X-ray diffraction patterns to show that the DNA molecule consisted of a double helix.

MAGNETISM

From navigating the globe
to deflecting deadly radiation,
magnetism impacts on all our lives

Nominated by **Tom Heap**, presenter of
BBC *Panorama* and *Costing the Earth*

On 13 March 1989, the lights went out over Quebec. For nine hours, 6 million people sat in the dark, some wondering if the incredible light show above their heads was the first strike of a much-feared nuclear attack. But the cause of the blackout wasn't the result of some earthly dispute; it had a more cosmic origin.

The Sun's atmosphere is so hot that it emits a constant stream of deadly radiation made up of charged particles – the so-called 'solar wind'. The protective blanket of the

"To me magnetism is one of the greatest discoveries of all time, because of its impact throughout history on everything from exploring our world to causing the fantastic light shows of the aurora – but also because of its relevance today in developing renewable technologies."
Tom Heap

Earth's magnetic field deflects most of these particles, but sometimes a few manage to sneak through, smashing against atoms of gas in the upper atmosphere to create the spectacular light show of the auroras. When the particles hit oxygen atoms these 'Northern Lights' glow red or green, and when they collide with nitrogen atoms they appear blue or purple.

Occasionally the Sun gets extra feisty and releases a massive belch of charged particles. If our planet is in the firing line of this 'coronal mass ejection' (CME), it can be deadly to astronauts and cause havoc on Earth. When a billion-tonne CME hits the magnetic field, the vibrations from the impact induce currents powerful enough to overload electrical circuits and even melt transformer components. In the March 1989 event in Quebec, the region took the full force of a CME, which wiped out the vast Hydro-Quebec power network.

Fortunately CMEs aren't frequent or strong enough to penetrate the Earth's magnetic field often, because, without this protective blanket, much of life on this planet would not survive the radiation onslaught.

Journey to the Centre of the Earth

The Earth's magnetic field extends several thousands of kilometres into space. Where the charged particles in the solar wind collide with the magnetic field, the magnetosphere forms. Its shape resembles iron filings around a bar magnet, but it is compressed on the daylight side of the Earth, where it is exposed to the solar wind, and extends out on the sheltered night-time side of the Earth. But what exactly is it that causes this magnetic field?

The Earth is made up of a number of different layers from the crust to the molten iron core. The 'dynamo theory' proposes that heat radiates from this intensely hot core, generating convection currents which 'stir' the molten iron, creating weak magnetic forces. Very occasionally – every few thousand to a million years – the magnetic field flips, so that magnetic north is at the South Pole. If our dynamo theory is correct, this could be because of turbulence in the iron core.

The Japanese research ship *Chikyu* is currently drilling what will be the world's deepest hole, 12 km down into the Earth's crust. But as the planet has a radius of over 6,000 km, no-one can prove whether the dynamo theory is correct until we actually journey to the centre of the Earth. All we do know for sure is that it's generated internally, not externally as once thought.

Before the seventeenth century, many people believed either that the Pole Star (Polaris) created the Earth's magnetic field or that there was a large magnetic island on the North Pole. Then, in 1600, William Gilbert, the physician to Elizabeth I, published *De Magnete*, a book all about magnetism. Gilbert based his ideas on experiments with a *terella*, a small magnetised model ball representing the Earth. He realised that the inside of the Earth must be magnetic and proposed iron as the substance responsible. This was a big breakthrough in our understanding of magnetism, because we were finally able to explain why a compass needle points north.

Navigating the World's Oceans

Before the magnetic compass was employed for navigation around AD 1100, sailors had to keep land within sight in order to navigate their way around the seas – or risk a watery death.

The first time a magnetic device was recorded as being used for orientation was a Chinese reference made in 1044 to a small iron model fish in a bowl of water. By 1120 the fish had evolved into a needle in water, and the proto-compass was born. In 1269 French scholar Pierre de Maricourt produced the first record of magnetism, writing at length about magnetic poles and forces, but also about compass needles. Then, by 1300, every seafaring mariner throughout Europe possessed a compass consisting of a pivoted magnetised needle housed inside a dry box.

"To me magnetism is the greatest discovery," says Heap, "because of its impact on navigation, enabling us to explore this world of ours."

But the magnetic compass only got sailors so far. They could tell where north was, give or take a bit, and by the end of the fifteenth century they knew that magnetic north wasn't quite the same as geographic north, so they would adjust their readings to account for the difference. Mariners could also work out how far north or south on the globe they were – the latitude – by measuring the angle of the Sun in the sky using an astrolabe or a sextant.

What they couldn't do was work out the longitude – how far east or west they were. Many a reef or rocky shore claimed the lives of those on board ships that foundered after straying off course. Frustrated captains had to sail north or south to the latitude of their final destination, and then sail directly east or west and hope for the best.

Tired of losing valuable cargo at sea, in 1714, the British parliament offered a £20,000 reward – a hefty sum in those days – to anyone who could come up with an ingenious way to work out longitude. Numerous people came up with astronomical solutions that required a high calibre telescope and a steady deck. But, in 1759, English clockmaker John Harrison finally cracked it.

Harrison's invention was a robust clock, called a 'chronometer'. Its spring-driven mechanism could cope with a ship's deck that pitched and rolled. It was so accurate that sailors could compare the time on the chronometer with another clock set to local time. By calculating the difference in time, they could then work out their longitude.

ELECTROMAGNETS

Inside any electric motor lies a fixed magnet, whose attracting and repelling forces turn the motor's coil, thereby generating electricity. This invention was a collaborative effort over many years by a number of different scientists and inventors, and much of modern-day technology relies on the electric motor. "The science of magnetism has had a whole new lease of life with the advent of renewable energy technology," says Heap.

The generator of a wind turbine uses electromagnets to turn the blades' motion into electrical power. The more efficient you can make magnets, the more efficiently you can generate electricity. "The latest wind turbine technology uses rare-earth metals, such as lanthanides, in the electromagnets," says Heap. "Not only are they more durable than, for example, iron which gradually loses its magnetism over time, but they are also more efficient."

The unique properties of rare-earth metals are the result of how their electrons are configured in each of their atoms. Their outer electron shells are similar to those of metals such as iron, but their inner shells include so-called '4f electrons'. These interact with the electric field of the central nucleus in ways that give the rare earths the special properties exploited in so many high-tech products.

Demand for these rare-earth metals is high, but China has a stranglehold on these vital materials, holding 97 per cent of the known global resource. "China is flexing its muscles," says Heap. "It has already threatened to pull its consumer tech export to Japan."

A resource war over rare-earth metals could be on its way.

The magnetic compass is just part of the history of navigating the seas and exploring our planet. But without it and other instruments, such as the sextant and the chronometer, explorers may never have been able to traverse the globe, making such accurate maps for future seafarers to follow.

ELECTROMAGNETISM

How a stray compass needle led to the dawn of a new way of thinking

Nominated by **Dr Marcus Chown**, author whose books include ***Quantum Theory Cannot Hurt You***

Many of the gadgets and gizmos that we use every day rely on electromagnetism to work. Without its discovery, there would be no TV to watch or speakers to listen to music. We have a Danish chemist to thank for its discovery.

Rumour has it that on 21 April 1820 Hans Christian Oersted was setting up some apparatus for a lecture when he noticed something strange going on with his compass: the needle was at an odd angle. He traced the cause to a nearby wire and realised that its electric current must be deflecting the compass needle. He'd discovered electromagnetism. A whole host of scientists picked up on this and labs around the world buzzed as people tried to find a use for it.

'James Clerk Maxwell is arguably the most important physicist between Newton and Einstein, as he was the first person to realise that the fundamental reality that underpins the world is totally unlike the familiar everyday world of our senses.'

Marcus Chown

When Frenchman André-Marie Ampère heard of Oersted's new breakthrough in September 1820, he immediately set about playing around with magnets and electricity. Just one week later he was in the French Academy of Sciences demonstrating his new discovery. He showed that two wires carrying an electric current can be made to attract or repel each other just like magnets, depending on which way the current is flowing in each wire.

But it was a bookbinder in London who made the ultimate breakthrough. Bored with his day job, Michael Faraday had been trying to get into the scientific world for a long time. Then, in 1821, he got his lucky break. When Humphry Davy from the Royal Institution temporarily blinded himself during an experiment and his assistant was fired for assaulting the instrument-maker, Davy needed a new one. Faraday was on hand – and keen.

Faraday spent many months tinkering with magnets and electricity. And, at last, he got his reward – creating a device that made an electric current rotate a wire around a magnet. Little did he know just how ubiquitous his rudimentary electric motor would one day become.

Solenoid

Ampère started playing around with Faraday's device. He found that winding wire into coils intensified the effect – the more coils of wire, the stronger the magnetic field when an electric current passed through the wire. And so the solenoid was born. Solenoids are now used in all sorts of everyday devices, including a car's locking system: locking and unlocking the car remotely changes the direction of the current in the coil, which in turn changes the direction of the magnetic fields.

Using Ampère's results, English physicist William Sturgeon created a very strong electromagnet by placing a bar of iron inside the solenoid, making it capable of lifting 4 kg. Then along came American scientist Joseph Henry, who, in 1830, made a super-strength one that could lift 340 kg, the weight of five people. The uses for such a strong magnet in technology were endless. While vital for the development of technology, there was no satisfactory theory to explain exactly what was going on. It would take a mathematical physicist from Scotland to rescue the situation. In so doing, he would unify electricity and magnetism into a single theory.

Maxwell's Equations

Faraday was a hero in the world of science in his day and was responsible for founding the Christmas Lectures at the Royal Institution in 1825. He claimed that electricity and magnetism worked along 'lines of force'. But that was as far as his non-mathematical brain could take him.

Scottish mathematical physicist James Clerk Maxwell was a talented scientist. He was educated at Edinburgh University and Trinity College, Cambridge, and became a professor at Aberdeen University at the tender age of just 25. When he moved to King's College in London, he started to work on Faraday's idea, trying to find out what was going on using mathematics. Initially, he tried to model the world according to how man viewed it. "Maxwell tried to explain how a magnet transmitted a force to a piece of metal by imagining tiny, invisible cogs turning in the space between the magnet and the metal," says Chown. "Finally, he gave up and threw away the cogs."

Gradually it dawned on him that he had to look at the world in another way. "Instead, he imagined ghostly electric and magnetic 'force fields' with no parallel in the everyday world permeating space," says Chown.

Maxwell had worked out that electricity and magnetism, as well as light, existed in wave form – electromagnetic waves. "It was a seismic break with the past," says Chown. "It liberated physics, and led to the quantum world where atoms can be in two places at once, the world of general relativity, and, later on, string theory."

Maxwell's set of four mathematical equations now govern everything we know about electromagnetism. And his discovery paved the way for a German physicist to reveal the existence of radio waves.

RADIO WAVES

How proving that electromagnetic waves exist opened the door to wireless communication

Nominated by **Iain Lobban**, Director GCHQ

David E. Hughes was a prolific inventor: he improved Edison's telephone transmitter and invented a crucial part of the metal detector. But he should have been best known for his radio-wave transmitter and detector, if only someone had actually listened to him.

In 1879, fifteen years after James Clerk Maxwell had predicted the existence of electromagnetic waves (see page 163), Hughes spotted some sparks from a transmitter circuit that seemed to be affecting an unconnected telephone system hundreds of metres away. Realising he

"Starting with the simple modulation of a carrier wave, Marconi's invention of the radio transmitter rapidly made it possible for individuals to communicate with each other wherever they were in the world."
Iain Lobban

was on to something, he asked for an audience with fellows at the Royal Society and demonstrated the effect. But they were unimpressed, dismissing it as a simple case of electromagnetic induction – when voltage is produced as a conductor moves through a magnetic field. He tried his luck at the Post Office too, but, once again, no-one was interested. So, Hughes's discovery slipped into oblivion.

Irish physicist George F. Fitzgerald is also often forgotten for his role in the discovery of radio waves. In 1883, he described how a current going back and forth through a conductor could create low-frequency long-wavelength electromagnetic waves – so-called 'radio waves'. Fitzgerald had the theory but lacked the evidence. Hughes had the evidence but lacked the theory. In 1888, a German scientist had both.

Hertz's Proof

Heinrich Hertz was a bit of a polyglot – learning Arabic and the Indian language Sanskrit – as well as being a bright spark in theoretical physics. It was while lecturing at the University of Karlsruhe that he made his most important discovery.

In 1888, Hertz managed to produce the radio waves that first Maxwell had envisioned then Fitzgerald had predicted. To create these radio waves, he developed a device with two antennae. A sudden blast of energy from a capacitor caused electrons oscillating in a long wire to generate radio waves from one antenna. When they hit the second antenna, these radio waves created another current.

Hertz had proved what others had only predicted – and for that, the unit used to describe the frequency of oscillations in radio waves is named after him. However, finding a commercial use for such a device fell to a Serbian-born engineer.

Tesla's Genius

Nikola Tesla was born to a priest and an illiterate mother in a Serbian village in what is now Croatia. Tesla was a genius with a photographic memory and an ability to build his inventions from the detailed pictures he'd conjured up in his head – no pen required. He led a tumultuous young life – cutting himself off from his family, suffering from illness and nervous breakdowns, city-hopping from Budapest to Paris, and eventually ending up in New York.

When Tesla arrived in the US, he had little more than a letter of recommendation from one of his former employers addressed to Thomas Edison (the Edison of light-bulb fame). The letter reportedly said, "I know of two great men and you are one of them; the other is this young man".

Tesla lived up to the letter, solving key electrical engineering problems for Edison's company before going on to form his own company. It was in the 1890s that he made his key contributions to radio communication, building antennae which were able to transmit high-frequency radio waves over vast distances – and creating a system of wireless telegraphy. But while he was a great inventor, Tesla was a poor businessman and ended his life as an impoverished outcast.

Going Wireless

After reading about Tesla's work, Italian inventor Guglielmo Marconi set about developing new components for the wireless telegraphy system, including a device that generated a varied electrical resistance – the so-called 'coherer' – which helped detect radio signals.

While the wireless systems of others were limited to a few hundred metres, over time Marconi was able to extend the range of the antennae and eventually transmit signals over hills. He realised that, with better equipment, there was no limit to how far signals could be transmitted, so he travelled to London to drum up funding. Where once the Post Office hadn't been interested in Hughes's radio wave discovery, this time round the chief electrical engineer saw the enormous potential of Marconi's long-range wireless telegraphy. With British backing, in 1897, Marconi's system succeeded in first sending Morse Code signals across Salisbury Plain, then across the Bristol Channel and finally, in 1901, a message travelled wirelessly across the Atlantic from Poldhu in Cornwall to Newfoundland. Cable telegraphy now had a competitor.

In 1904, British physicist John A. Fleming developed the first valve diode, which changed oscillating signals received by an antenna back into a direct current that could be detected more easily. But the biggest breakthrough in wireless transmission happened in the early 1900s.

AM and FM

Wireless transmission had been hampered for decades by the fact that it was difficult to transmit low-frequency waves of sound at high frequencies. To get round this, Canadian inventor Reginald Fessenden invented the 'heterodyne system', which worked by mixing the sound wave with a higher-frequency carrier wave so that the signal had the same strength as the carrier. This came to be known as 'amplitude modulation' (AM). But interference was still a problem. So, in the 1930s, American inventor Edwin H. Armstrong developed another method: frequency modulation (FM).

The invention of modulation through AM and FM systems has meant that audio signals can be transmitted clearly over long distances. Radio waves and the ability to modulate them have dramatically changed how we we receive information, and, crucially, how we communicate.

HOW RADIO WAVES ARE TRANSMITTED

To transmit radio waves efficiently, lower frequency sound waves are mixed with carrier waves that have a constant strength (amplitude) and higher frequency. In amplitude modulation (AM), the varying amplitude of the sound signal modulates the carrier signal amplitude but not the frequency (see below). In frequency modulation (FM), the varying frequency of the sound signal varies the carrier signal frequency but not the amplitude.

Radio waves begin when sound enters a microphone in waves of air pressure fluctuations. Different pressures in the sound waves are converted to an electrical sound signal, which is then 'ramped up' (or amplified) before feeding through to the transmitter, where the signal forces electrons inside the antenna to move, producing electromagnetic waves.

These waves are then picked up by an aerial, linked to a radio with a tuning circuit that is tuned in to a certain frequency. A transistor then splits the modulated signal into the carrier signal and sound signal, which is then boosted by the amplifier and carried through to the speaker. The signal causes the diaphragm (a thin, flexible disk covering the front of the speaker) to vibrate, producing sound waves.

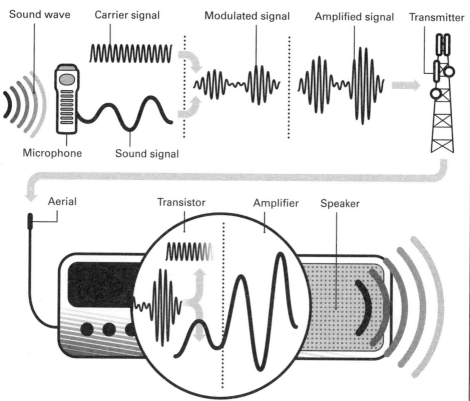

Sound wave   Carrier signal   Modulated signal   Amplified signal   Transmitter

Microphone   Sound signal

Aerial   Transistor   Amplifier   Speaker

MICROWAVES

The discovery that changed wars – and meal times

Nominated by **Professor Colin Blakemore**, professor of neuroscience at Oxford and Warwick universities, and former chief executive of the British Medical Research Council

Few modern technologies have a history as extraordinary as that of microwaves, which have helped win wars and cooked countless TV dinners, but first found fame as a 'death ray'.

From the very earliest days of research into radio waves, scientists knew they could carry energy wirelessly. In the 1920s, reports of aircraft crashing near powerful radio transmitters prompted a British inventor named Harry Grindell Matthews to claim he had a device emanating invisible rays which could kill vermin and stop engines tens of metres away.

The British military expressed interest, but evidence for the effectiveness of the 'death ray' amounted to little more than some amateur movie footage of doubtful authenticity.

While Matthews and his claims quickly fell into obscurity, fascination with the idea persisted. In 1935, a committee of UK defence scientists asked Scottish engineer Robert Watson-Watt to investigate whether radio waves were capable of becoming such a weapon. Watson-Watt showed that the idea was completely unfeasible, as the amount of radio power needed was immense. But he added that it was possible that radio waves might have another use, as an air-raid warning system.

The ability of radio waves to bounce off metal objects was another feature noted soon after their discovery. Watson-Watt and his colleagues exploited it to create a system they called RADAR – RAdio Detection And Ranging. The technology received a huge boost in 1940 with the invention by physicists at the University of Birmingham of the cavity magnetron, a device that generated short-wave radio waves known as microwaves.

With wavelengths of around a few centimetres, these proved ideal for detecting aircraft and even the periscopes of submerged submarines. As such, radar played a key role in the Allied victory in World War II. It remains vital today in everything from air-traffic control to detecting storms.

A chance discovery during the development of radar led to another application for microwave technology. In 1946, engineer Dr Percy Seymour of the US Raytheon company was working with a magnetron when he noticed a chocolate bar in his pocket had started to melt. Within months, his company had patented the idea of using a magnetron to cook food.

Today, microwave ovens can be found in millions of homes, but they're not the most ubiquitous form of this incredibly versatile technology. That claim to fame goes to the mobile phone, where the short wavelength, and thus high frequency and information-carrying capacity, is used to carry data, speech and images.

Technology that detects microwaves has proved no less revolutionary. In the 1950s, astronomers discovered that hydrogen atoms – the most common form of ordinary matter in the Universe – emit a characteristic microwave signal. Known as the '21 cm line', it travels through deep space and even dust, allowing astronomers to glimpse the otherwise invisible structure in our own galaxy and millions of others.

Most dramatically of all, in the mid-1960s, yet another chance event allowed microwave technology to gain credit for one of the most famous of all scientific discoveries. Engineers at Bell Labs in the US were puzzled by the background 'hiss' being picked up by a huge microwave aerial designed to receive satellite communications. They tried everything to get rid of it, even scraping out pigeon droppings – without success. Two engineers, Arno Penzias and Robert Wilson, noticed that the hiss was the same no matter which way the aerial was pointing. This suggested the microwaves were coming from deep space (see page 101).

They learned that astronomers had in fact already predicted the existence of the microwaves. Their origin was nothing less than the Big Bang, the cataclysmic explosion which gave birth to the Universe around 14 billion years ago. Originally incredibly intense, the radiation from the Big Bang had been diluted and stretched by the expansion of the Universe until it turned into a feeble hiss of relatively long-wavelength microwaves.

Today, those self-same cosmic microwaves can be picked up on any detuned TV set – a fitting demonstration of the power and wonder of microwave technology.

The Chemical World

CHEMICAL BONDS

The key to understanding the structure of substances

Nominated by **Professor Tom Welton**, professor of sustainable chemistry and head of deparment at Imperial College London

Way back around 400 BC, the Greek philosopher Democritus was sitting at home one day when the smell of freshly baked bread wafted through the air. Democritus reputedly had a eureka moment when he realised that minute particles must have floated between the loaf of bread and his nostrils. He named these particles 'atoms'. If his assumption was correct, then solid objects, Democritus reasoned, must be held together in some way, maybe like the 'hook and eyes' commonly found in items of clothing. While many of Democritus's ideas were way ahead of their time, his theory on how atoms hold together was very wide of the mark.

"Chemical bonds between atoms make the material world that we inhabit and experience possible."
Tom Walton

Atoms hold together by the movement of electrons, but the electrons can actually be shared or transferred in a number of different ways.

TYPES OF CHEMICAL BOND

Atoms contain a nucleus surrounded by shells of electrons. There are 2 electrons in the first shell, 8 in the second and 18 in the third. In order to be less reactive and more stable, atoms often share electrons and fill up their shells.

Covalent bonds

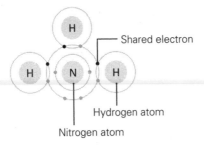

Shared electron

Hydrogen atom

Nitrogen atom

Covalent Bonds

In covalent bonds, atoms share electrons with other atoms in order to reach a stable state. For example, in ammonia (NH_3), covalent bonds exist between a nitrogen atom and 3 hydrogen atoms.

Hydrogen Bonds

Weak electrical attractions form between hydrogen atoms and oxygen, and nitrogen. In water, the oxygen atom takes more than its fair share of the electrons and so has a slight negative charge, while the two hydrogen atoms have a slight positive charge, so an electrostatic attraction forms between the atoms.

Hydrogen bonds

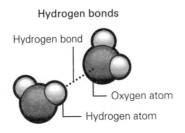

Hydrogen bond

Oxygen atom

Hydrogen atom

Ionic Bonds

Ions are electrically charged particles formed when atoms lose or gain electrons – giving them either a positive or negative charge. Unlike covalent bonding, the bonding electron is not shared but instead transferred. Electrons from the outer shell of one atom move to fill a gap in the outer electron shell of another atom. For example, in sodium chloride (common salt), a sodium atom gives up one electron to become a sodium ion, and a

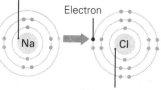

Ionic bonds

Sodium atom (becomes sodium ion once electron is lost)

Electron

Na   Cl

Chlorine atom (becomes chloride ion once electron is gained)

chlorine atom gains that electron to become a chloride ion. Electrostatic attraction holds the two ions together.

Metallic Bonds

Solid metals – or alloys – have a rigid crystal lattice. Metal atoms lose their electrons from their outer shell, which then slosh around in a sea of loose electrons that move freely through the lattice. The rigid lattice makes a metal strong, yet the free electrons mean it can conduct electricity and heat easily.

Metallic bonds

Rigid crystal lattice

Electron (moving freely)

Van der Waals Forces

These are much weaker types of chemical bonds where no electrons are shared or transferred, but simply a sort of 'stickiness' exists between layers of atoms. For example, in graphite, sheets of carbon are held together by Van der Waals forces.

Van der Waals forces

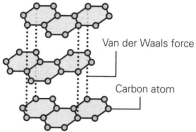

Van der Waals force

Carbon atom

There are in fact five major types of chemical bond (see pages 178–179), which vary in their strength of attraction. The history of how each was discovered spans a number of centuries and involves a number of players.

In his famous book *Opticks*, published in 1704, Isaac Newton looked into the puzzle of how atoms were attached to one another. He realised there must be some kind of force holding them together, but he never discovered exactly how that worked. It took a chemist from Östergötland in Sweden to come up with a plausible theory.

Jöns Jakob Berzelius started out as a physician, but got drawn into the world of chemistry. In 1819, he developed a theory that there must be some sort of electromagnetic force attracting atoms to one another. By the mid-1800s, a group of chemists had developed this idea further into the 'valence theory', whereby the attraction was due to positive and negative 'poles' – electrons and protons. One of the chemists, August Friedrich Kekulé, looked into how many bonds atoms could form, concluding in 1857 that carbon could form four, so had a valency of four. He also discovered that a benzene molecule (C_6H_6) forms a symmetrical ring of carbon atoms. He claimed the idea came to him in a dream in which a snake was eating its own tail.

The American chemist Gilbert N. Lewis developed the concept of the electron-pair bond in 1916, whereby two atoms can share up to six electrons. This became known as 'covalent bonding'. That same year, the German physicist Walther Kossel proposed a different theory – that electrons are completely transferred between atoms. This is 'ionic bonding'.

The more chemists investigated, the more they started to realise that covalent and ionic bonds didn't account for the structure of all substances. In fact, the Dutch scientist Johannes Diderik van der Waals had hit on a clever idea many years before. In 1873, his observations of gases and their low boiling points suggested that some weaker chemical bond must exist between atoms. Now known as 'Van der Waals forces', the bond between atoms in these sorts of substances is about 100 times weaker than ionic or covalent forces – more like the stickiness in Pritt Stick, as opposed to superstrength glue.

It's now known that Van der Waals forces exist in various forms: three attractive and one repulsive. An example of an attractive inter-molecular force is the hydrogen bonds between water molecules, which are absolutely crucial for life on Earth. The repulsive example is the hydrophobic force, which causes water to 'bead' on the waxy surface of leaves.

The discovery of the various different types of chemical bond has been vital for human development and the evolution of technology. As Welton points out: "By understanding the way atoms bond, we have learned how to manipulate them and create our modern world with all of its wonderful man-made materials that have made our lives so much healthier and easier than those of our forebears."

DID YOU KNOW?

Diamond is such an incredibly hard substance because of its strong covalent bonds between its carbon atoms that form a rigid 3D network, each atom being equidistant from its neighbouring carbon atoms.

OXYGEN

The discovery of the gas of life

Nominated by **Dr Hal Sosabowski**, principal lecturer in chemistry and pharmaceutical sciences at the University of Brighton and presenter

Frenchman Antoine Lavoisier was rich, very rich. He'd inherited a fortune, but also fed off a corrupt government system as a 'tax farmer'. For a fee, the government let men like Lavoisier collect taxes on their behalf. But surprisingly there was nothing he and his wife Marie liked to do more on a Sunday than hole up in their laboratory for hours on end, tinkering with different chemical substances.

Lavoisier was a great chemist, but also

"The discovery of oxygen was a massive breakthrough not only because it led to our understanding of chemical reactions but also because it is the very essence of what keeps us alive. We live on a planet with just the right amount of oxygen in the atmosphere: 20.8 per cent. Below 17 per cent there wouldn't be enough to burn carbohydrates in our cells, but above 25 per cent oxygen is toxic."
Hal Sosabowski

DID YOU KNOW?

- At −180°C oxygen is a blue liquid. Few elements are coloured.

- Liquid oxygen is paramagnetic: if suspended in a test tube on a piece of string, a strong magnet will move the test tube.

- Oxygen reacts with every single element, apart from the inert gases.

- Oxygen is toxic above 25 per cent. The high oxygen tension in some incubators has been known to blind premature babies.

- Above 25 per cent oxygen levels, all organics are flammable; the fire in the first manned *Apollo* spacecraft couldn't be put out because the oxygen in the cabin was enriched.

- Oxygen is manufactured in some stars: hydrogen is burnt to helium, to carbon and to oxygen, and at 1 billion degrees it then becomes silicon, phosphorus and sulphur.

- One hundred million tonnes of oxygen are produced industrially every year, normally by fractional distillation of liquid air, which involves cooling air down until it becomes a liquid, and then warming it up and collecting the various fractions (gases) as they boil at their different boiling points.

- Ozone is toxic to humans, damaging the lungs, but without the ozone layer in the atmosphere, UV would damage life on Earth.

- The static sparks seen when taking off a jumper in the dark produce minute amounts of ozone.

an ambitious man. At a dinner party in October 1774, he saw his opportunity not only to make more cash but also to become a science celebrity. One of the dinner guests was English chemist Joseph Priestley. Foolishly, Priestley related how the previous August he'd discovered something quite incredible.

Heating a substance called mercuric oxide over a very intense flame created an unusual type of air that made a candle burn vigorously. When Priestley took a deep sniff of the unusual air, it made him feel somewhat light-headed. Little did he know that he'd produced the gas that we now call oxygen.

Priestley called it 'dephlogisticated air'. But he wasn't the first to discover the gas – a Swedish chemist Carl Scheele had heated mercuric oxide two years before with the same results. But Scheele never saw a benefit to his discovery. Nor did Priestley – at least not at first. He later realised he'd missed a trick.

Lavoisier repeated Priestley's experiment and later recombined the mercury and oxygen to produce the original mercuric oxide powder. The more he played around with substances, the more he realised that they could be amalgamated and then broken apart, again and again. Lavoisier had made an incredible find – he'd discovered how chemical reactions work.

When Lavoisier renamed 'dephlogisticated air' as 'oxygen', Priestley was miffed to say the least. But while both Scheele and Priestley should be credited for discovering oxygen, you have to hand it to Lavoisier for understanding how simple elements can combine to form different compounds – and that matter can neither be created nor destroyed. Indeed, it was this discovery that set the scene for the breakthrough in understanding the very nature of matter.

THE BUNSEN BURNER

The invention that unlocked the secrets of the elements and led to the creation of the Periodic Table

Nominated by **Dr John Emsley**, University of Cambridge, author of books such as *Molecules of Murder* and *Nature's Building Blocks*

For more than a century the Bunsen burner was the chief tool of the chemist, and was to be found on every lab bench. Indeed, it played a critical role in the evolution of chemistry.

Before the 1850s, traditional burner lamps used to heat samples in labs were inefficient with a feeble flame. When the well-respected chemist Robert Bunsen moved to Heidelberg in Germany in 1852, the university was just starting to build a chemistry lab that was hooked up to the city's new coal-gas street lighting. Perfect timing for Bunsen to play around with the design of a new type of burner lamp that used the building's gas supply.

Bunsen asked the university mechanic Peter Desaga to create a prototype that generated a very hot but sootless flame. Desaga set to work, incorporating a clever piece of design that would radically change the face of chemistry.

"The crucial part of the design was the little vent at the bottom of the burner, which allowed you to regulate the air coming in, mixing air with the gas before it was burnt," says Emsley. "This produced an intense, very hot flame."

Well-known chemist Michael Faraday and not so well-known gas engineer R. W. Elsner had come up with similar designs before, but Bunsen was the one to mass produce it. By the time the new lab opened in 1855, Desaga had built 50 such burners for the students. But Bunsen's invention wasn't just useful for student science: it paved the way for a radical breakthrough in the discovery of new elements.

"Bunsen's invention allowed an element to be identified via its atomic spectrum," says Emsley. The light emitted in a hot flame could be observed using a device called a 'spectroscope' (see page 189). Every element produces a unique fingerprint of spectral lines. If you observed a new line or lines, then you knew you had a new element, and in this way several were discovered in the 1860s. Bunsen himself discovered caesium and rubidium by means of spectroscopy.

As more and more scientists began to use the Bunsen burner, more and more elements were discovered. Around 60 elements were known in Bunsen's day, but by the end of the nineteenth century the total was nearer 90. "But back in 1860 the known elements all looked a bit like an unorganised stamp collection," says Emsley. That too was soon to change, thanks to a Russian chemist by the name of Dmitri Mendeleev. He brought order to the chaos.

Russian Revolution

Mendeleev was born in a small village near Tobolsk in Siberia. Throughout his life he was on the move, from the frozen wastelands of Siberia to St Petersburg to the Black Sea coast, where he became a science master at one of the oldest public schools in Ukraine. In the late 1850s and early 1860s he worked on the spectroscope at Heidelberg, before returning to St Petersburg. It was while he was a professor in the city that he came up with his most revolutionary piece of work – the Periodic Table.

Frustrated by the lack of order between all the known elements, Mendeleev attempted to classify each according to its chemical properties. He wrote the properties down on a set of cards. Rumour has it that he started to arrange the cards as you would in a game of Patience. And very gradually he noticed a pattern forming, but within the pattern were some glaring gaps.

By comparing the properties of the known elements bookending these gaps, Mendeleev was able to predict as yet undiscovered elements. Other chemists, such as John Newlands and Lothar Meyer, had spotted patterns in the elements, but Mendeleev was ahead of his time in predicting elements that would one day reveal themselves in a lab. Mendeleev presented his work to the Russian Chemical Society in 1869, showing how the elements could be grouped in such an arrangement according to their atomic weights.

The Periodic Table is now the poster boy of chemistry. In fact, most schools around the world have a picture of it pinned up somewhere. Over the years the Periodic Table has bulged and fattened as new elements have been discovered, but it has maintained the basic structure that Mendeleev incredibly devised way back in 1869. It currently contains 118 elements.

"The Periodic Table had to expand as new elements were discovered and then to be rearranged according to atomic theory," says Emsley. "But it was remarkable that Mendeleev arrived at his table before anyone knew about the nature of atoms."

An element's identity is defined by the number of protons in its nucleus, and its chemistry by its electrons. "Mendeleev had the confidence to just do it using simple, logical deduction," says Emsley. "It was his Periodic Table which set chemistry on a firm theoretical foundation."

DID YOU KNOW?

In 1843 a sample of an arsenic compound (then called 'cacodyl cyanide', aka dimethyl arsenic cyanide) exploded in Bunsen's face. He was then severely affected by inhalation of the arsenic fumes and was ill for several weeks, as well as losing the sight in one of his eyes, which had been pierced by a splinter of glass.

THE SPECTROSCOPE

How splitting light revealed the structure of the atom and new elements

Nominated by **Professor Pat Roche**, head of the astrophysics department at the University of Oxford

The different colours you see in a rainbow make up the visible spectrum. Philosopher and Franciscan friar Roger Bacon was the first to notice the visible spectrum, while studying the range of colours displayed in a rainbow, but it wasn't until the 1670s that the great Isaac Newton was the first to actually describe the colours in white light as a 'spectrum'.

"The discovery of spectroscopy was so important because it unlocked the structure of atoms and was widely used in astronomy, physics and earth sciences for analysing materials, detecting pollutants and monitoring the Earth's atmosphere."
Pat Roche

After hacking a hole in his window shutters, he positioned a prism so that the narrow beam of light from outside hit the prism at an angle, splitting the light into its constituent colours, which were displayed

HOW A SPECTROSCOPE WORKS

Chemical elements and compounds absorb – and emit – light at specific wavelengths. This depends on the 'energy gap' between their various shells of electrons. Plotted on a graph with light intensity on one axis and wavelength on another, the absorption or emission shows up as troughs and peaks. The patterns formed reveal the properties of the element or compound.

A spectroscope receives and disperses incoming light, or other radiant energy, into a spectrum, usually by diffraction or refraction. Spectroscopes vary in design depending on what is being analysed. This is how a typical spectroscope works:

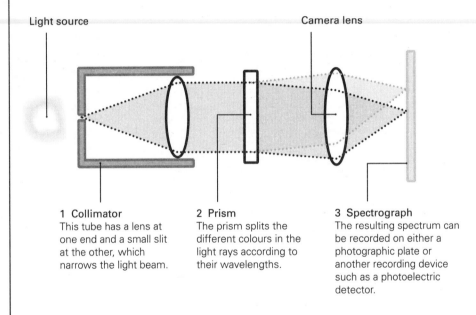

Light source

Camera lens

1 Collimator
This tube has a lens at one end and a small slit at the other, which narrows the light beam.

2 Prism
The prism splits the different colours in the light rays according to their wavelengths.

3 Spectrograph
The resulting spectrum can be recorded on either a photographic plate or another recording device such as a photoelectric detector.

on the wall opposite. He found that a second prism recombined them into white light.

Ever the self-promoter, who knew how to spin a good yarn to intrigue people, Newton claimed he could see seven colours – violet, indigo, blue, green, yellow, orange and red. Not many people can distinguish violet from indigo, but his claim certainly adds to his legendary status.

Applications for this technique didn't emerge until the mid-1800s.

Chemist Robert Bunsen moved to Heidelberg in Germany in 1852. Taking advantage of the new coal-gas street lighting system in town, he commissioned the university mechanic to create an extra hot burner lamp – so hot that it would change the face of chemistry.

When heated, any substance emits a distinct line in an otherwise uniform and continuous spectrum. This varies between substances, which means that each has its own unique signature, or 'spectral line'. When cooled, each substance also absorbs light at the same wavelength at which it emitted the spectral line.

Using his Bunsen burner (see page 185), Robert Bunsen worked alongside Gustav Robert Kirchhoff, discovering new elements such as caesium and rubidium. Analysis of the patterns of spectral lines in elements, such as hydrogen, also led to the development of the Bohr model of the atom (see page 125) and revealed many other atomic properties. But scientists also started to realise the potential of spectroscopy in astronomy.

ASTRONOMICAL SPECTROSCOPY

How the spectroscope revealed our expanding Universe

Nominated by **Professor Richard Dawkins**, evolutionary biologist and emeritus fellow of New College, Oxford; his books include *The Selfish Gene* and *The God Delusion*.

"Confounding the doubts of philosophers, the Fraunhofer fingerprint lines in a star's spectrum betray the elements that it contains. Every galaxy's redshift tells us the speed with which it is receding, and this dates the origin of all things. Through the spectroscope, we know that time, space and reality began 13.7 billion years ago."

Richard Dawkins

Few practising doctors are also successful scientists and businessmen. But William Hyde Wollaston was no ordinary doctor: he was the secretary of the Royal Society from 1804 to 1816; he discovered the elements palladium and rhodium; he made a fortune from developing a method to process platinum ore commercially; and he invented the camera lucida (although this has been disputed). Wollaston's biggest contribution to science, however, could well be his discovery in 1802 of mysterious dark features in the Sun's electromagnetic spectrum.

It wasn't until the German physicist Joseph von Fraunhofer invented the spectroscope in 1814, and more carefully repeated Wollaston's experiments, that he identified the strange features as dark lines in the Sun's electromagnetic spectrum, which became known as 'Fraunhofer lines'. Forty-five years later, Robert Bunsen and his colleague Gustav Robert Kirchhoff showed that different Fraunhofer lines are emitted by different elements and substances (see page 189).

Soon, spectroscopic discoveries weren't just limited to our Solar System. From his private observatory in Tulse Hill, South London, amateur astronomer William Huggins and his wife observed all sorts of celestial objects. Huggins was the first person to work out that stars are composed mainly of hydrogen; 73 per cent of our Sun's mass, for example, is hydrogen. On 29 August 1864, he became the first to capture the spectrum of a planetary nebula – a glowing shell of gas ejected from a dying star.

The next big breakthrough in astronomical spectroscopy occurred in the early 1900s. Using Vesto Slipher's results, Edwin Hubble showed that nebulae were rushing away from the Earth (see page 101). What Hubble had discovered was the so-called 'redshift' of galaxies. The closer an approaching ambulance comes towards you, the higher pitched the siren sounds. After it's passed by, the pitch sounds lower. Likewise, as a moving star or galaxy draws nearer, its light is shifted to the blue end of the spectrum; while the faster it recedes, the more its light is shifted towards the red end.

The telltale redshift that Hubble saw showed that the galaxies were racing away from the Earth, proving that the Universe was expanding. This astounding breakthrough enabled us to calculate the date of the Big Bang – and, crucially, prove that our Universe is 13.7 billion years old.

ARTIFICIAL DYES

The discovery that artificially coloured our world

Nominated by **Aidan Laverty**, editor of TV series **Horizon**

People have used natural dyes for thousands of years. Dyed flax fibres from 36,000 years ago have been discovered in a prehistoric cave in what is now modern-day Georgia. In India, dyeing has been carried out for more than 5,000 years using dyes obtained from natural objects like roots, berries, bark, leaves and wood.

In ancient Egypt, the deep blue pigment that coloured the eyes of the magnificent gold funery mask of Tutankhamen was made from the semi-precious stone lapis lazuli. It was the same colour that was later used by Renaissance artists to paint skies. In Roman times, one of the commonest ways to make the colour purple was from the natural dyes found in the glandular mucus of the humble sea snail. "Emperors may have looked an imposing purple, but up close they could smell of a left-over seafood supper," says Laverty.

> *"William Perkin is the unsung father of fashion and design. He added colour to our world through the artificial dyes that he pioneered."*
>
> Aidan Laverty

What both these colours have in common is that they were naturally occurring, expensive and therefore exclusive. Until, that is, a certain British chemist came along with the first artificially produced colour – mauve.

William Heny Perkin was born in 1838 in the East End of London, the son of a local carpenter. He showed an early interest in science and at the age of 15 entered the Royal College of Chemistry in London. At the age of 18, during his Easter break, Perkin discovered the first synthetic organic chemical dye: mauveine. As with so many scientific breakthroughs, his discovery was a complete accident after an experiment went wrong.

DID YOU KNOW?

Any colour you see is not within the object itself, but is in fact the colour of the light reflected from the object.

At the time, Perkin was working as an assistant to the grandly named August Wilhelm von Hofmann. Hofmann was intent on trying to create a synthetic form of the naturally occurring yet expensive substance quinine, which was heavily in demand as a treatment for malaria. Perkin carried out a series of experiments in his makeshift lab at his home in Cable Street during the Easter holidays. Completely by accident, he created a crude mixture from aniline which, when extracted with alcohol, produced a strange substance that was an intense purple colour.

Excited by what he'd discovered, Perkin carried out more experiments with two friends in his garden shed at home. In order not to get into trouble for not focusing on the quinine research, he kept it a secret from Hofmann.

On 26 August 1856, Perkin was granted a patent for "a new colouring matter for dyeing with a lilac or purple colour stuffs of silk, cotton, wool, or other materials". Realising its enormous commercial potential, he opened a dyeworks factory in London. The colour was also known as 'aniline purple', but by 1859 most people knew it as mauve. It became the in-colour in fashionable circles, worn by the likes of Queen Victoria.

At the relatively youthful age of 36, Perkin had made enough cash to sell his company, devoting the rest of his life to research. Since Perkin's discovery, thousands of synthetic dyes have been developed, eventually leading to the chemical industry we know today.

Perkin's discovery marked the moment when the synthetic started to take over from the natural on an industrial scale.

PLASTICS

The rise of synthetic polymers

Nominated by **Professor Donal Bradley**, Lee-Lucas Professor of Experimental Physics at Imperial College London

We use nearly 300,000 tonnes of plastic bottles in the UK every year. In the US, the plastics industry employs over a million people and is the third biggest manufacturing industry. Every year, enough plastic is produced in the US to shrink-wrap the state of Texas.

Strong, lightweight, yet cheap to manufacture – plastics are one of the few materials that can claim to be all of these things. In addition, many plastics are also poor electrical conductors, making them ideal for insulating electrical equipment. It's not hard to see why they are so prevalent in the modern world.

The first man-made plastic was 'parkesine', named after its Birmingham-born inventor Alexander Parkes. He showcased it at the Great International Exhibition in London in 1862. Four years later, he founded the Parkesine Company, planning to mass-produce the material, but his cost-cutting ways came back to bite him as the product wasn't of sufficient quality. One of Parkes's colleagues, Daniel Spill, improved on parkesine to form xylonite before this became registered as celluloid in 1870. It wasn't until the 1920s that the structure of plastics (polymers) was understood.

German chemist Hermann Staudinger was investigating the structure of natural materials like cellulose, starch and proteins, when he came up with the idea of polymers (*polymeros* is Greek for 'many units'). In his paper in 1920, he suggested that they were giant molecules made up of chains of covalently linked repeating units (see page 177). At the time, chemists thought plastic was made up of bunches of small molecules, not long chains. But Staudinger's breakthrough ushered in a materials revolution that led to such household names as styrofoam, nylon, polythene, neoprene and teflon.

The type of atoms and the way in which they bond together to form the repeated links of a polymer chain define its molecular structure, which in turn controls many of its bulk properties. "Some molecular structures are well suited to electrical insulation," says Bradley. "Others, for example that of Kevlar, provide exceptional toughness, ideal for modern-day body armour. The way in which the polymer chains pack and interact is also very important, for example the chains in Kevlar armour are highly ordered."

The molecular nature of polymers means that they typically have strong covalent bonding along the chain, but the van der Waals forces keeping the neighbouring chains together are relatively weak, roughly one-twentieth of the strength of a covalent bond (see page 177). "This helps processing," says Bradley, "because the chains can be relatively easily separated, by heat or a solvent, and then reassembled into new shapes, for instance conforming to a curved surface or the elaborate features of a mould."

The earliest commercial polymers used natural cellulose and were developed to replace expensive natural materials, for example celluloid

HOW PLASTICS ARE MADE AND RECYCLED

Plastics are made by the 'polymerization', or linking together, of small molecules (monomers) typically derived from crude oil, gas and coal, but more recently also from bioproducts such as sugar-cane bioethanol. The word 'plastic' is derived from the Greek *plastikos*, meaning 'deformable, malleable or mouldable'. There are about 40 different types of plastics manufactured, the most common being high-density polyethylene (HDPE), polyethyleneterepthalate (PET) and polyvinylchloride (PVC).

Tiny pellets of plastic that look a bit like fish eggs, known as 'mermaid's tears', are often found in large numbers on UK beaches. Around 100,000 marine animals are thought to be killed each year from ingesting plastic or becoming tangled in it (although, for comparison, figures for the number killed by other types of marine pollution are hard to come by). As there is a lot of concern about marine plastic waste, research is being done to try to fully understand its effects and how they might be remedied, by, for example, using biodegradable plastics and safer processing additives.

Most plastics are non-biodegradable, meaning that they will take hundreds or even thousands of years to decompose if left in landfill. Large-scale recycling now takes place in the UK at material reclamation facilities – a machine recognises and sorts the different types of plastics that are then squashed and chopped into small chips that are cleaned, melted and sold on for reprocessing.

Recycled plastic can be turned into all sorts of items, from carpets to electrical fittings. It takes about 25 plastic bottles to produce a new fleece jacket, and recycling just one plastic bottle saves enough energy to power a 60 W lightbulb for six hours.

as an ivory substitute and rayon for silk. Then, in 1909, the first fully synthetic plastic, Bakelite, was created. Over the years, a myriad of other synthetic plastics have been developed, from polyethylene to polystyrene. But now a new generation of plastics is sweeping the world's labs.

"There's a lot of excitement around recent developments in 'conjugated polymers', that have all the traditional attributes of plastics, but can also carry a current like semiconductors and metals," says Bradley. "It means that gadgets like large flat-screen TVs might be made using methods more familiar in newspaper printing. Other prospects include lighting and solar energy conversion, and indeed the combination of both in a solar light. It opens up a whole new vista for many of the devices that populate our technologically intensive environment."

The Living World

THE MICROSCOPE

The device that revealed the invisible world around us

Nominated by **Dame Athene Donald**, professor of experimental physics at the University of Cambridge and chair of the Royal Society Education Committee

For centuries, people have been able to zoom in on the microscopic world using lenses. Records suggest that ancient Romans used crude magnifying glasses to read tiny or faint print. But it was the *Book of Optics* by the great Arab inventor Alhazen, published in 1021, that was hugely influential on the development of early types of glasses, for correcting long- and short-sightedness.

Tracking down who invented the first microscope in the late sixteenth

"The development of the microscope in the seventeenth century opened up the world of the very small, permitting the discovery of cells and, as microscope manufacture improved with the production of better lenses, details within cells too. Now, almost every branch of science relies to some extent on one or more of the different variants of microscopy."

Athene Donald

century isn't easy. Spectacle-maker Hans Janssen and his son Zacharias from Middelburg in the Netherlands are known to have put a single, tiny lens inside a brass tube. Realising that if one lens magnified by a small amount, two lenses would zoom in even further, they added another lens to create a so-called 'compound miscroscope'. Then, another Dutch spectacle-maker, Hans Lippershey, who was virtually a neighbour of the Jansenns, designed the first proto-telescope with convex and concave lenses at either end of a tube.

When Italian polymath Galileo Galilei heard about Lippershey's device, he set about building his own version of the telescope, realising how useful it woud be to see impending Ottoman Turk invasions. Galileo also built a microscope and called it 'little eye' which, in 1625, German botanist Giovanni Faber renamed 'microscope'.

It was Robert Hooke who created the first compound microscope that really worked well. An oil lamp produced the light source, which was concentrated onto an object by a large lens. Inside the microscope were four tubes, containing an object lens and a middle lens. By moving the microscope closer or further from the object, Hooke could find the correct focus to view it clearly.

And it was Hooke who played a crucial role in using his microscope to discover the miniature world.

CELL THEORY

Discovering the units that make up every single living creature

Nominated by **Dr Adam Rutherford**, science writer,
video editor at the journal **Nature**, and presenter of BBC series
The Cell and **Genome**

Peering through his home-made microscope, Dutch cloth merchant Antonie van Leeuwenhoek gasped in surprise. He'd brought some water back from a local pond and while inspecting it saw tiny creatures moving around, so minute that they were invisible to the naked eye. In everything that he analysed under his microscope, from the blood of a tadpole to his own semen, Leeuwenhoek

"All life is made of cells, and all cells can only come from other cells. This describes the entirety of biology."
Adam Rutherford

discovered more and more of these minuscule creatures, which he named 'animalcules'.

Being a draper, Leeuwenhoek needed strong magnifying lenses to assess the quality of the cloth he bought. So he made powerful lenses no bigger than a small raindrop for his home-made microscope, and the strong magnification revealed a hidden miniature world which no-one else had seen.

Intent on getting some expert opinion on what these creatures could be, Leeuwenhoek in 1674 sent a package containing a letter and drawings of the animalcules to the Royal Society in London. In it he reported all his findings, including his discovery of the animacules in human semen: 'Sometimes more than a thousand in an amount of material the size of a grain of sand.'

When the package landed on Robert Hooke's desk, he tried to zoom in on Leeuwenhoek's hidden world by using his own microscope to examine water from the Thames. But he found nothing.

With no scientific background and poor English, Leeuwenhoek's claims were initally ignored by the scientific establishment. But realising that the Dutchman might be on to something big (or small), Hooke persevered in his attempt to create ever-more powerful lenses. And, at last, his microscope revealed Leeuwenhoek's secret world – creatures moving around beneath the lens, more blurred than the Dutchman's detailed drawings, but there nonetheless.

In 1680, Leeuwenhoek's contribution to science was finally acknowledged when he was elected Fellow of the Royal Society. Although no-one yet could explain how this microscopic world made up the whole human body, Leeuwenhoek's discovery was momentous.

"It was only through the development of better lenses by Leeuwenhoek that scientists began to see things that they hadn't seen before," says Rutherford. "That was a huge breakthrough."

Buddies and Betrayal

It was Hooke, though, who first coined the term 'cell'. In 1665, almost a decade before he'd received Leeuwenhoek's package, he published

Micrographia, describing the world he'd discovered through his own rudimentary microscope. One particular object he analysed was cork bark, slicing it open to reveal a regular structure of interlocking units. As they reminded him of the box-like cells in monasteries, he reportedly called the units 'cells'.

But it wasn't until the nineteenth century that cell theory emerged. At the same time as German botanist Johann Moldenhawer proved plant cells were discrete units by gently teasing them apart at the walls, a botanist and a biologist were developing the idea of cell theory. Matthias Schleiden and Theodor Schwann realised that, although there were differences between the plant and animal cells they were studying, the cells were essentially the same basic building blocks for all organisms.

"Until that point there had been very little crossover between plant and animal biology," says Rutherford. "There was this notion that plants were made of plant stuff and animals made of meaty stuff. Schleiden and Schwann suggested that living material didn't simply form spontaneously, as previously thought, but that all living tissues are made of cells, and cells are the smallest unit of life."

This wasn't proved until the mid-1800s though, when the story gets quite juicy. Polish scientist Robert Remak was studying red blood cells in chick eggs when he spotted something intriguing – a cell splitting into two.

As Remak was a Jew, he wasn't allowed a permanent place at the University of Berlin, so he showed his results to his mentor and then friend German physician Rudolf Virchow. When Virchow realised the enormity of what Remak had discovered, he published the results in a textbook, gaining all the credit. Virchow went on to become a major player in science, politics and social reform, but, inevitably, the friendship ended.

"Remak was the first person to see a cell budding into two," says Rutherford. "Virchow had essentially stolen his work."

Syn-cells

We now know that each of us is made up of anywhere between 10 and 100 trillion cells. This is an incredible feat considering we develop from just two cells typically fusing in the mother's fallopian tubes to form a zygote, whose cells then divide as the embryo develops.

The importance of cell theory and its impact on today's progress is vast. "The essential carrier for DNA, and the transmitter of all life for 4 billion years, has been the cell. But now the cell is entering a new era, where we can get DNA to do exactly what we want, when we want," says Rutherford.

This is a bold new era: the era of artificial cells and synthetic life.

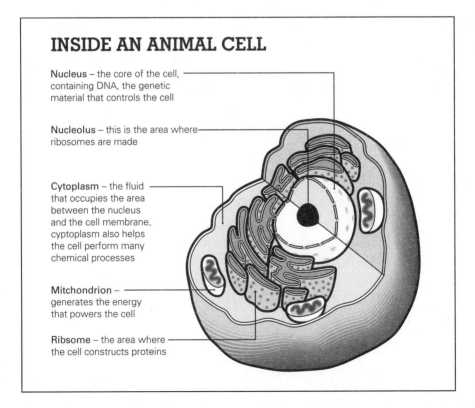

INSIDE AN ANIMAL CELL

Nucleus – the core of the cell, containing DNA, the genetic material that controls the cell

Nucleolus – this is the area where ribosomes are made

Cytoplasm – the fluid that occupies the area between the nucleus and the cell membrane, cyptoplasm also helps the cell perform many chemical processes

Mitchondrion – generates the energy that powers the cell

Ribosome – the area where the cell constructs proteins

EMBRYO DEVELOPMENT

The discovery of human fertilisation

Nominated by **Professor Lewis Wolpert**, emeritus professor of cell and developmental biology at University College London, whose books include *How We Live and Why We Die*

If someone told you that a piece of cheese could give birth to a mouse you'd probably laugh out loud. Ridiculous as this now sounds, for hundreds of years most scientists believed that living creatures gradually came to life from non-living matter: an idea known as 'spontaneous generation'.

That was until the nineteenth century, when German botanist Johann Moldenhawer showed that plant cells were separate units, by peeling them apart, and Matthias Schleiden and Theodor Schwann proposed cell theory (see page 207). But if someone had just

"Along with the belief in the divine creation of the world was the belief that all embryos had existed from the beginning of the world. This belief was dispelled at the end of the nineteenth century with the discovery that even human embryos develop from a single cell: the fertilised egg."

Lewis Wolpert

listened to an Italian Catholic-priest-cum-scientist a century or so before, this discovery might have happened a lot earlier.

Lazzaro Spallanzani was always questioning the fashionable theory of the time; ever controversial, he was never one to accept something as fact. When he took over as director of the Natural History Unit in Pavia in 1768, he continued to lecture enthusiastically and also to travel extensively abroad. He would return from his trips laden with wonders from the natural world. It was these collections that were to inform his theory of animal reproduction – and challenge the idea of 'spontaneous generation' of life.

Spallanzani had a hunch that both semen and an ovum were needed to produce an embryo. He conducted a series of slightly bizarre experiments. To prove his hypothesis that sperm and the ovum had to come into contact, he dressed some male frogs up in taffeta shorts, while others wore nothing. Sure enough, only the female frogs that mixed with the short-less male frogs got pregnant.

In other experiments, Spallanzani showed that a paintbrush dipped in frog semen and smeared on unfertilised eggs caused tadpoles to develop, and if semen were placed in a very fine sieve eggs weren't fertilised, as the sperm couldn't get through the fine mesh. Then, in 1777, Spallanzani was the first person to perform *in vitro* fertilisation by articially inseminating a dog.

Spallanzani died in 1799 in Pavia, succumbing to bladder cancer. After his death, his bladder was removed and studied by colleagues, before being put on public display in the museum in Pavia, where it still remains today.

Half a century later, in 1852, Henry Nelson claimed to have watched the fertilisation of the *Ascaris* worm through his microscope. But in this and other observations of sperm penetrating eggs in sea urchins and starfishes, the fertilisation took place outside the creatures' bodies. It wasn't until 1826 that German zoologist Karl Ernst von Baer located eggs in the ovaries of a dog. Then, in the early nineteenth century, Swiss physicist Jean-Louis Prévost and French chemist Jean-Baptiste Dumas discovered embryos in the fallopian tube, indicating that this organ is the site for fertilisation and proving that human embryos develop from a single cell, the fertilised egg.

CLASSIFICATION OF SPECIES

Making sense of the millions of species on planet Earth

Nominated by **Professor Danielle Schreve**, professor of quaternary science at Royal Holloway, University of London

In the 'great chain of being' described in the fourth and fifth centuries AD, God was at the top, then came angels, humans, animals, plants and, finally, non-living matter such as rocks and minerals. The idea had originated in ancient Greece from philosophers including Aristotle, but Christian theology had moulded it into that lineage, which survived for many hundreds of years. At each step of this ladder, there were more 'rungs', so humans were broken down into lords at the top and serfs at the bottom.

It was a brave move for someone to suggest a different system for classifying the different beings on God's Earth. But one man realised there was a need to bring order to the vast array of species on the planet.

The exoticism of Jamaica lured Irish doctor Hans Sloane to take up the position of personal physician to the governor of the island in 1687.

While there, this amateur botanist collected all sorts of creatures and plants. But when his employer died suddenly 18 months later, Sloane loaded his collection on to a boat and returned home.

Some of these natural treasures he stored away. Others he made use of, such as the bean of the cacao tree, which when mixed with milk created a tasty drink that Sloane patented. That same recipe was eventually bought by the Cadbury family.

Over the years, Sloane amassed a vast array of species, which in 1742 he moved to his new manor house in Chelsea, which later became the Chelsea Physic Garden. Local streets including Hans Crescent and Sloane Square are named after him. When he died, he bequeathed his hoard of natural wonders to the nation and it became the founding collection at the British Museum. Sloane left a legacy, including a classification system that he'd devised in Jamaica. But some didn't think the system watertight.

John Ray had befriended Francis Willoughby while they both studied at Cambridge University. After graduating the pair toured Europe, studying the natural world around them. Ray divided the 18,000 flora and fauna he discovered along the way into how they looked and the terrain they inhabited. His big contribution to science was identifying how one species isn't born from the seed of another. A Swedish naturalist took Ray's classification system a step further.

Nils Linnaeus was the first in his family to ditch the traditional patronymic naming system, where the father's name becomes the stem of the surname. Instead he adopted the surname Linnaeus, after the giant linden tree on the family land. His son, Carl Linnaeus, also changed his name later in life becoming Carl von Linné. It wasn't too egotistical an act: he was ennobled in 1761 for the enormous part he played in classifying the world around us.

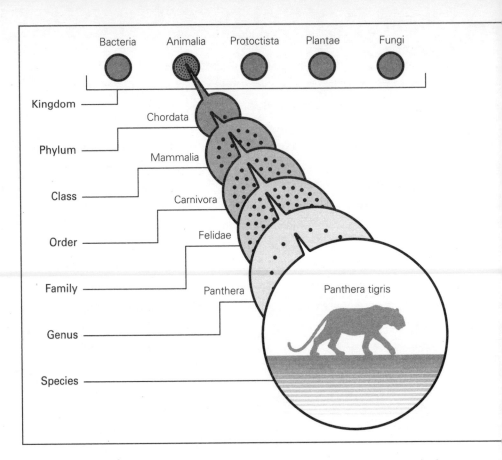

Kingdom — Bacteria · Animalia · Protoctista · Plantae · Fungi

Phylum — Chordata

Class — Mammalia

Order — Carnivora

Family — Felidae

Genus — Panthera

Species — Panthera tigris

Apparently, as a boy, Linnaeus learnt Latin before Swedish, which could be why in the binomial system he created, each creature or plant had a Latin name made up of two words (genus and species) to establish biological relationships, based on shared physical characteristics. In his enormous multivolume encyclopaedia, *Systema Naturae*, he divided life into three kingdoms, subdivided into classes, then orders, genera and, finally, species.

"It was an astonishingly bold step for Linnaeus when he attempted to classify every organism on Earth," says Schreve.

Kingdom – The highest rank recognised by Linnaeus. He originally named just two kingdoms of life (plants and animals), but bacteria, protoctista and fungi were later recognised.

Phylum – Large differences split animals into the different phyla, for example the physical make-up of the body, or the way in which the embryo develops.

Class – This category can include a wide range of organisms, distinguished by less fundamental biological differences, such as skin coverings (scales in fish, feathers in birds).

Order – Further differences are classified in this category, such as the distinctive skeletal structures or tooth arrangement.

Family – Yet more detailed differences are evident to further separate groups into recognisable 'families'.

Genus – This level of classification separates different groups of the same family – so recognising the differences between big and small cats, for example.

Species – The end result of the classification table, individual species as we know and recognise them today.

Linnaeus travelled extensively around Europe during his life – even organising a trip to Lapland. While in Holland, he was curator of the botanical garden at Heemstede, and on a visit to London in 1736 he met up with none other than Hans Sloane, whose Chelsea Physic Garden was rearranged according to Linnaeus's classification system.

Linnaeus died in 1783 and his vast collection passed to his widow Sara. Upon her death, a young medical student, James Edward Smith, bought all 14,000 plants, 3,198 insects, 1,564 shells and thousands of letters and books – for a mere £1,000. A massive admirer of Linnaeus, Smith founded the Linnean Society in 1788.

Although there have been subsequent revisions of some of Linnaeus's original groupings, in particular drawing on new evidence from molecular genetics, they still remain an incredibly useful tool in biology. Schreve says: "His revolutionary system transcends international language barriers and remains a fundamental part of science today."

CONVERGENT EVOLUTION

How animals' bodies – and minds – evolve to be closer

Nominated by **Professor Nicola Clayton**, FRS, professor of comparative cognition at Cambridge University

You could forgive eighteenth-century scientists like Carl Linnaeus for thinking that echidnas and hedgehogs came from the same family. Indeed, their spiny bodies make them look very similar. Through genetic tests, we now know that hedgehogs are insectivores, while echidnas are monotremes, or egg-laying mammals.

Their physiological similarity is the result of convergent evolution: when two organisms from completely different species evolve similar traits. So, for example, bats, insects, birds and, once upon a time, pterodactyls, all evolved wings. Convergent evolution takes place because similar environments pose similar challenges to an organism, and natural selection 'picks' from one of a finite number of solutions to the problem. In the case of wings, there was only really this option for successful flight.

Littering the natural world are numerous other examples of convergent evolution: whales, shrews and bats all use echo location; anteaters and aardvarks have long sticky tongues; and sharks and dolphins both have what's known as 'countershading', a form of camouflage where the upper side is darker than the lower, blending the animal into its surroundings when light shines on it from above. Many of these species live in completely different parts of the world and evolved down totally separate lineages, but when challanged by similar conditions they evolved similar physiological characteristics.

"Convergence can occur in very distantly related groups, where similarities arise as a result of adaptation to similar selection pressures," says Clayton. "The more distantly related the groups, the stronger the case for convergence."

Convergent evolution also seems to occur with mental characteristics. Charles Darwin suggested that abilities like memory and cognition are subject to evolution by natural selection in much the same way as morphological traits. For many years, it was assumed that it was only our close ancestors, the primates, that had cognitive abilities.

Clayton and her colleague Nathan Emery believe that a few other more distantly related animals have these abilities too: "We think that cognitive abilities evolved through convergence in crows and apes because they faced the same social and ecological challenges, despite having very different brain architectures."

NATURAL SELECTION

The evolution of evolution

Nominated by **Dr Alice Roberts**, Director of Anatomy at the NHS Severn Deanery School of Surgery, and author and presenter

Born with twelve fingers and twelve toes, Robert Chambers as a child underwent corrective surgery and had two of each amputated. But the operation left him lame. This could have been the undoing of a less intelligent man, but Chambers immersed himself in the world of books, and, alongside his brother, set up a publishing company in Edinburgh. The enterprise was a huge success, and gave Chambers the chance to try his luck at writing.

Vestiges of the Natural History of Creation rolled off the printing press in 1844, but no-one knew the name of the author. It stirred up a hornet's nest of controversy. But what had got Victorian society's knickers in such a twist? An attack on God. The book pieced together various threads of ideas that were murmuring dissent in different corners of science. It combined new theories on the cosmos, geology and the fossil record, and went so far as to suggest there was no need for a grand creator. In other words, God had got the elbow and the Church didn't like it.

But the book rapidly became a best-seller – even Queen Victoria and Prince Albert read it. The author remained anonymous, however, and it wasn't until after his death in 1871 that Chambers was confirmed as having written it.

Darwin wasn't a fan of *Vestiges*. Secretly he was intrigued by the book's radical ideas, but his religious convictions and work as a naturalist didn't sit well alongside it. "His geology strikes me as bad and his zoology even worse," he said. As Darwin watched the book's savage basting by the critics, he realised the need to collect extensive physical evidence to support his own ideas on evolution; otherwise he, too, would get a verbal lashing.

Darwinism

Born in 1809, Darwin lived a tumultuous early life. He left a medical career at Edinburgh University, mainly due to not being able to stomach nineteenth-century surgical methods. He turned to the Church, studying theology at Cambridge, but failed to pass his exams, having been distracted by the natural world around him. His obsession with beetles and

"To me, natural selection is the greatest idea of all time. It seems as if man has been around forever. But in the geological timescale, against the background of the vast antiquity of the Earth, we've been here just a brief moment. We like to think we represent the pinnacle of life on Earth. But the theory of evolution through natural selection knocks us off that pedestal. We're a tiny twig on the immense tree of life, and our evolution, our existence as a species today, was not an inevitable, foregone conclusion. We're lucky to be here and, because of Darwin and Wallace, we know it!"

Alice Roberts

the like landed him a place on HMS *Beagle*, which set sail in December 1831 for a round-the-world trip. It was Darwin's discoveries on that journey that anchored the *Beagle* in the annals of history and sculpted our understanding of evolution.

On his travels, Darwin encountered creatures great and small. In the wildlife mecca of the Galapagos islands, he came face to face with giant tortoises and iguanas, and collected all sorts of animals from bugs to birds.

After five years, he returned home, laden with exotic discoveries, and set about piecing these together around a radical idea of evolution that was forming in his mind. It was at one meeting with his friend the ornithologist John Gould that the penny dropped. Gould spotted that the birds Darwin had classified as mocking birds were in fact from the same family as the finches the naturalist had collected. Referring back to his detailed notes, Darwin found that the different finch beak shapes coincided with the availability of different food sources. Just as in the book he'd read by Charles Lyell claiming that the Earth was very gradually changing, Darwin realised that life had gradually evolved into different species over the aeons of time due to different environmental pressures.

But it was a piece of work by British economist Thomas Malthus, *An Essay on the Principle of Population*, that cemented Darwin's radical idea – a theory he called 'natural selection'. In his work, Malthus discussed how a growing human population would eventually result in competition for finite resources. Darwin realised that the same would happen in the animal kingdom. If a fox didn't have enough food to eat, then it would fight for the last scraps with another fox, and the strongest would live to tell the tale. So competition would result in the 'survival of the fittest'. Darwin never actually coined this phrase, but it has become synonymous with 'natural selection', where different species evolve from the same initial group.

While speciation can take anything in the region of tens of thousands of years to a few million, evolution can happen much more quickly. A good example of speed-evolution in action is in the cichlid fishes that inhabit the clear waters of Lake Victoria. They have diverged into a number of completely distinct species in just 15,000 years. Quicker still, physical characteristics have been found to evolve in a matter of decades. In 1930, about 1 per cent of African elephants did not have tusks. As these elephants were not hunted by poachers, not having any lucrative ivory, they survived to pass on the genes for this tuskless trait. In Africa, tuskless elephants now make up as much as 38 per cent of the population. This is bad news for elephants, however, as they need their tusks for debarking trees, digging and fighting.

Zoönomia

Darwin was by no means the first to form a theory of how life evolved on Earth. His own grandfather, Erasmus, had published *Zoönomia* in 1794, which put forward the idea that higher forms of life had evolved from more primitive ones.

A French soldier with the unwieldy name Jean-Baptiste Pierre Antoine de Monet, Chevalier de la Marck, aka Lamarck, came up with the first fully developed theory of evolution. In the early 1800s, Lamarck suggested that acquired characteristics could be passed on to offspring. So, the neck of a giraffe stretching to reach for high leaves would become longer during its lifetime, and its calf would in turn be born with a longer neck.

Thanks to Darwin's theory of natural selection, we now know Lamarck was wrong (although he was on the right track by claiming the environment can influence our physiology: see page 245). It took many years for Darwin's idea of natural selection to be coaxed out of him. When his beloved daughter Annie died in 1851, his religious beliefs finally crumbled. Then a timely letter from the naturalist and explorer Alfred Russel Wallace, who had come to the same conclusions as Darwin, triggered him to speak up.

In 1858, Darwin and Wallace published a joint paper, describing natural selection. A year later, Darwin published his renowned book *On the Origin of Species*, which generated a huge wave of controversy. Although it was written for the general reader, Darwin was eminent enough for it to be taken seriously by the scientific community. Then, just three years later, evidence was unearthed that cemented Darwinism.

Dinobird

In a Bavarian quarry in 1861, a slab of rock was split open to reveal the skeleton of an animal with feathered wings. The feathers suggested it was some ancient bird, but the claws on the wing bone and the length of bony tail looked just like a reptile's. A local doctor bought the fossil, then pawned it off for a pricey £700 to the superintendent of the natural history collection at the British Museum. Eventually it fell into the hands of naturalist Thomas Henry Huxley, who was the first to spot the fossil for what it was. It had been named *Archaeopteryx*, meaning 'ancient feather', but that didn't do it justice. This fossil was not simply an ancient bird, but a link between reptiles and modern birds – and, crucially, precious evidence of the chain of evolution.

In Darwin's day, no-one knew the exact mechanism behind natural selection. In the case of finches, we now know that the gene calmodulin influences beak shape, but back then genetics wasn't even a glint in the eye of evolutionary theory. It wasn't until an Augustinian monk, Gregor Mendel, started experimenting on pea plants that the mechanism for inheritance of characteristics was unveiled.

MENDELIAN INHERITANCE

How an Augustinian monk discovered genes

Nominated by **Dr Michael Mosley**, author, and BBC producer and presenter of programmes including *The Story of Science*

Gregor Mendel wasn't your average monk. During his childhood, he worked as a gardener and studied bee-keeping. At 18, he moved to study at the University of Olomouc for three years, after which his physics teacher suggested he enter the Augustinian Abbey in Brno, in what is now the Czech Republic. Eight years later, he was back studying, but this time at the University of Vienna, where his professor of physics was Christian Doppler, who gave his name to the 'Doppler shift'. When Mendel returned to the abbey in Brno in 1853, it's not surprising that he couldn't leave science alone. Not only did he teach at the abbey, where he also studied astronomy and meteorology, but he bred bees and, crucially, pea plants.

Many before Mendel knew that characteristics were inherited. In 1745, French natural philosopher Pierre Maupertius claimed that offspring were made up of parts from every bit of their parents' bodies. Charles Darwin, of course, knew that favourable characteristics in the parents would serve the offspring well if they inherited them. But Darwin had never heard of Mendel's work, so no-one had worked out the exact mechanism behind natural selection.

Discovering Genes

Mendel bred peas to try to discover how to create better hybrids that would be hardier crops. At the time, people believed that when, say, a purple and white flower were crossed, it would produce a blended light-violet colour – an average of the two parents. But when Mendel analysed the characteristics of pea plants, it gradually dawned on him that specific characteristics could be inherited as one – often the offspring pea plants had either pure white or purple flowers. Mendel surmised that units of heredity inside the plants that he called 'factors' – now known as 'genes' – were dictating which colour the flower would be.

Crucially, Mendel realised that sometimes recessive genes were 'masked' by dominant genes, and so the recessive trait didn't always show up in the offspring. However, these recessive genes would still be passed on to the offspring, and the characteristic would skip a generation before reappearing in the next.

Mendel also discovered that organisms carry two variants of the same gene – known as 'alleles' – but pass on just one to each offspring. Mendel formed a Law of Independent Assortment, too. This stated that different characteristics are inherited independently of one another. For example, even if seed shape and seed colour were inherited together, they would not necessarily both appear in the offspring as they were in the parents. Each individual's set of alleles is known as their 'genotype', and

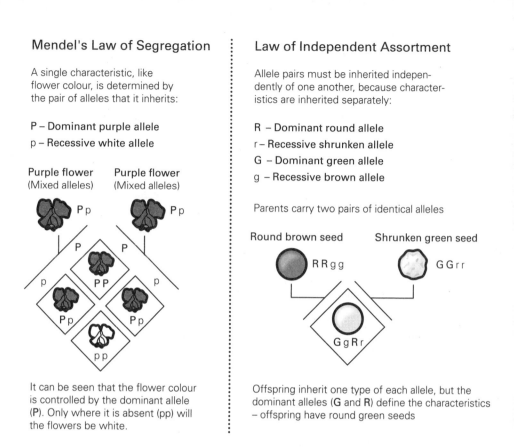

Mendel's Law of Segregation

A single characteristic, like flower colour, is determined by the pair of alleles that it inherits:

P – Dominant purple allele

p – Recessive white allele

Purple flower
(Mixed alleles)

Purple flower
(Mixed alleles)

P p

P p

P

P

p

PP

p

Pp

Pp

pp

It can be seen that the flower colour is controlled by the dominant allele (P). Only where it is absent (pp) will the flowers be white.

Law of Independent Assortment

Allele pairs must be inherited independently of one another, because characteristics are inherited separately:

R – Dominant round allele

r – Recessive shrunken allele

G – Dominant green allele

g – Recessive brown allele

Parents carry two pairs of identical alleles

Round brown seed

Shrunken green seed

RRgg

GGrr

GgRr

Offspring inherit one type of each allele, but the dominant alleles (G and R) define the characteristics – offspring have round green seeds

the visible effect of those alleles, in terms of an organism's physiology, is called the 'phenotype'.

Sadly, when Mendel took over as abbot of the abbey in Brno in 1868, he abandoned his scientific experiments – and his work eventually slipped into obscurity. When he died in 1884, his successor burned his papers.

Gaining Credit, at Last

Fortunately, Mendel's laws were rediscovered by the botanists Hugo de Vries and Carl Correns, although at the time both were unaware of the monk's prior discovery. De Vries had worked out that dominant and recessive genes exist, and why certain traits skip a generation. Meanwhile, despite Correns having corresponded with Mendel before his death, incredibly when he published his results of experiments on hawkweed, he didn't actually notice the connection of the monk's experiments with Darwin's theory of natural selection.

In 1900, Mendel was finally recognised for his groundbreaking discoveries and a century of genetic discoveries followed. "Gregor Mendel is one of the great romantic figures in the history of science – a humble monk toiling away in obscurity uncovers a great truth and is utterly ignored," says Mosley. "The reality does not quite live up to the legend. Mendel had no real understanding of what it was he had discovered and his data was so suspiciously accurate that he was later accused by a leading statistician (R.A. Fisher) of having manipulated his findings. Yet, in the end, none of this mattered, because what Mendel did in his abbey was truly groundbreaking and laid the foundations of modern genetics."

CHROMOSOME THEORY

Revealing the process of inheritance

Nominated by **Professor Kim Nasmyth**, professor of biochemistry and head of department at Oxford University

Once it was understood that embryos were descended from a single cell, namely the egg, it was clear that cell division powered the hereditary process. The key question then was how the paternal contribution was made. Observation of fertilisation showed that sperm entered eggs and created a male pronucleus that later combined with the female pronucleus, either by fusion or following the first nuclear division, a process now known as 'mitosis'. This suggested that it was the nucleus that carried the hereditary material. But how did the nuclei proliferate?

There was initially a furious debate over whether nuclear division was direct, in other words they grew and divided like cells, or indirect, undergoing division only after their constituents formed nuclear threads that contain the coiled DNA (now called 'chromosomes'), which then split in mitosis. This debate was settled in favour of the indirect mechansim by a talented German biologist.

Fire salamanders have a distinctive black skin speckled with yellow spots and stripes. As the name suggests, toxins secreted through the skin can be deadly to predators and also to humans. This didn't put off German biologist Walther Flemming, though. In the 1850s, biologists had started using a new type of stain known as 'aniline' to study cell structure. Flemming was using it to research cell division in fire salamanders, which are ideal candidates because of their large chromosomes.

In Flemming's day, biologists had no concept of what chromosomes were, so Flemming didn't quite realise what he'd discovered when he found that a certain structure in the salamander's gill cell absorbed more dye than the rest of it. In a paper published in 1878, Flemming called this structure 'chromatin'. He spent many hours bent over his microscope studying various different species, including sea urchins, recording how the chromatin gradually thickened and then moved in the cell-division process of mitosis.

Chromatin and chromosomes are basically the same thing, the difference being that chromosomes describe the condensed DNA that are about to divide in mitosis. Another German scientist, Heinrich Waldeyer-Hartz, named these 'chromosomes' in 1888, but it was Flemming who described the main stages of mitosis (see page 233), noting how the chromosomes initially line up in the centre of the cell and then move to different sides. He realised that the genetic material doubled prior to division, but didn't spot that each chromosome was made up of two identical pieces of genetic material – what we now call 'chromatids'.

Flemming knew absolutely nothing of the work of the Augustinian monk Gregor Mendel and the law of inheritance that he'd discovered some thirty years earlier while breeding pea plants (see page 225), but then neither did anyone else at the time, including two influential botanists.

Dutch botanist Hugo de Vries was awarded the Darwin Medal in 1906 for his work on inheritance. While studying the evening primrose, he discovered that inheritance was due to particles he called 'pangenes', later shortened to 'genes'. Meanwhile, German botanist Carl Correns had also discovered how inheritance works, while researching hawkweed. Remarkably, Correns didn't spot the significance of Mendel's work, despite having corresponded with him before Mendel's death. Later on, when the monk's work was rediscovered, both botanists acknowledged he'd beaten them to it.

Blissfully Unaware of Mendel

Two other biologists took their work a step further. While studying fertilisation in sea urchins and worms, Theodor Boveri discovered that the different chromosomes dictate how an embryo forms, because they carry different information, and that an embryo had to inherit a complete (haploid) set to undergo any semblance of normal development. This was a vital idea that in a sense was the discovery of the genome.

All this had taken place by 1902 in the complete absence of any awareness of Mendel's work. So, genetics sadly had no part in this phase of the chromosome theory of inheritance. The big breakthrough eventually came in 1902 when, upon rediscovery of Mendel's work, Walter Sutton and, independently, Boveri suggested that chromosomes were the carriers of Mendel's genetic determinants.

Sutton's experiments on grasshoppers showed that chromosomes occur in matching maternal and paternal pairs which separate during 'meiosis', the other form of cell division, where genetically different cells are produced, in contrast to the identical cells produced in mitosis.

Some scientists refused to accept this 'Boveri–Sutton chromosome theory'. "This could be for two reasons," says Nasmyth. "First, there were those who wanted a theory of inheritance couched in the language of physics and chemistry, and the notion that something was visible in the light microscope was tantamount to a resurrection of the homunculus hypothesis. In other words, it was inconceivable that something as big as a chromosome could be a molecule."

"Another reason for the theory's initial frosty reception is that it could not explain why (mitotic) chromosomes looked very similar in different cell types, even though they were presumed to be giving them very different instructions. This was a good point, but was a consequence not of the theory being wrong but rather due to it being incomplete. The solution to this paradox only started to emerge much later on with the work of François Jacob and Jacques Monod in 1961."

This concrete evidence was the final nail in the coffin, despite American biologist Thomas Hunt Morgan having proved back in the early 1900s that chromosomes were indeed the carriers of Mendel's genetic determinants (see page 234).

The debate on inheritance was over. "The 'big bang' was bringing together Mendelian inheritance and chromosome theory," says Nasmyth. "While the discovery of the structure of DNA was important, it was a postscript to the bigger breakthrough – chromosome theory."

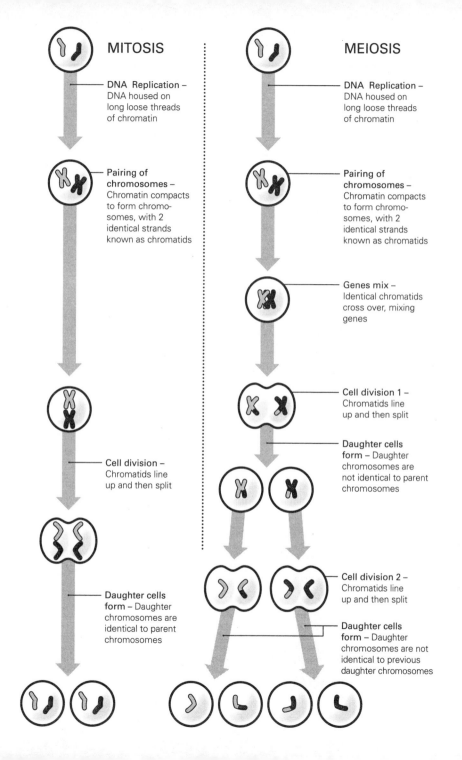

MITOSIS

DNA Replication – DNA housed on long loose threads of chromatin

Pairing of chromosomes – Chromatin compacts to form chromosomes, with 2 identical strands known as chromatids

Cell division – Chromatids line up and then split

Daughter cells form – Daughter chromosomes are identical to parent chromosomes

MEIOSIS

DNA Replication – DNA housed on long loose threads of chromatin

Pairing of chromosomes – Chromatin compacts to form chromosomes, with 2 identical strands known as chromatids

Genes mix – Identical chromatids cross over, mixing genes

Cell division 1 – Chromatids line up and then split

Daughter cells form – Daughter chromosomes are not identical to parent chromosomes

Cell division 2 – Chromatids line up and then split

Daughter cells form – Daughter chromosomes are not identical to previous daughter chromosomes

GENETIC RECOMBINATION

The discovery that led to the mapping of the human genome

Nominated by **Dame Kay Davies**, FRS, Dr Lee's Professor of Anatomy, director of the MRC Functional Genomics Unit and head of the Department of Physiology, Anatomy and Genetics at Oxford University

"Genetic recombination allowed the first genetic maps to be constructed. It explained the inheritance of traits and led on to the mapping of genes, as well as the determination of the chromosomal location of disease in humans."

Kay Davies

In his famous 'Fly Room' at Columbia University in the US, biologist Thomas Hunt Morgan bred fruit flies, *Drosophila melanogaster*. It was while researching the chromosome theory of heredity that, in 1911, Morgan spotted a strange phenomenon. He noticed that 'linked' traits, such as eye colour and gender, were always expressed in the same fly, suggesting that their genes stuck together during meiosis – cell division where non-identical daughter cells are produced (see page 229).

The bizarre thing was that other traits didn't show any noticeable linkage. After some time spent pondering this anomaly, Morgan realised that the chromosomes must be crossing over, and, where the crossover occurred, information and hence traits were shared.

We now know that this 'recombination' occurs during what is known as the 'prophase' in meiosis, when the two identical chromosome threads (chromatids) cross over, mixing genes. Morgan also noticed that the distance between two genes on a chromosome dictated how much linkage there was between them. A bit like beads that sit together on a necklace wound round the neck a number of times, two genes crossing over are more likely to be inherited together than two genes located at opposite ends of the chromosome.

With this in mind, undergraduate Alfred Henry Sturtevant, who was working with Morgan at Columbia University, realised it would be possible to map out the genes on any chromosome. Geneticists started constructing maps of all sorts of animal genomes – and, eventually, the human genome.

THE STRUCTURE OF DNA

Unravelling the
double helix

Nominated by **Dr Francis Collins**, Director of
US National Institutes of Health, and former head of
the Human Genome Project

Friedrich Miescher needed blood. He was a doctor, intent on studying the nature of disease, and for that he needed vast amounts of infection-fighting white blood cells. But where do you go if you need to find lots of the stuff? A war zone.

At Tübingen Castle in Germany in 1868, Miescher set up shop, collecting the blood-soaked bandages from injured soldiers who were fighting a war with Prussia nearby. He used the enzyme pepsin, which he'd extracted from the stomach of local pigs, to break down the walls of the white blood cells and isolate the cell nuclei. But inside each nucleus, he found something that completely baffled him: the chemical element phosphorus.

Previously, scientists believed the nucleus to be a protein. But phosphorus doesn't exist in proteins. Yet here it was, at the very heart of each cell. As Miescher broadened his search to animals, every cell he studied contained the element. Unwittingly, the doctor had uncovered a substance new to science, which he called 'nuclein'.

But 60 years came and went, with no-one interested in Miescher's insignificant nuclein. Then, in 1928, while looking into the cause of pneumonia, microbiologist Fred Griffith stumbled across the significance of nuclein. Working with two different strains of bacteria, he found that, after injecting them into mice, one strain killed the creatures but the other didn't. Griffith also found that if he mixed the lethal and harmless strains, the latter would often end up killing the mice. Sadly, he never got to the bottom of the mystery, as World War II intervened, and Griffith was killed in an air raid.

Nobel Snub

Across the Atlantic in New York, another microbiologist researching pneumonia found out about Griffith's results. Oswald Avery set about pulling apart bacteria to unravel what exactly caused the change that Griffith had uncovered. In the end, he found it – nuclein, or what we now call deoxyribonucleic acid (DNA). If Avery stripped out the DNA of each bacteria, the mice didn't die.

Avery's momentous discovery is so often overlooked in the annals of science. After he published his results in 1944, his boss didn't spot the huge significance of such a discovery, and even appealed to the Nobel Prize committee to ensure that Avery never received the award he surely deserved. But Avery wasn't alone in being snubbed by the Nobel bigwigs.

By the 1950s, biologists were increasingly convinced that it was DNA that played host to the long-sought genes believed to control the functioning of cells. However, exactly how this apparently simple molecule could pull off such a feat was still far from clear. Understanding that required a detailed knowledge of the structure of DNA, and that in turn required the use of the esoteric techniques of so-called 'X-ray crystallography'. At King's College London, a team of experts began the quest, among them Rosalind Franklin.

In X-ray crystallography, beams of X-rays are fired at crystals of the target compound. The regular atomic structure of the target crystal produces a 'diffraction' pattern on a photographic film that is characteristic of the target material. By using this technique, Franklin was planning to work out the 3D structure of DNA, captured in X-ray diffraction patterns on over 100 photographic plates. One of them was known as 'Photo 51'.

In 'Photo 51', Franklin could clearly see an 'X' in the middle of the image. This 'X' indicates that the X-rays passed through a spiralling structure, and that this structure was made of two such spirals – a 'double helix'. Franklin was so close to discovering the actual structure of DNA, but she wanted proof of the idea forming in her mind, so she waited to publish her results, intent on amassing more results.

Double Helix

New Zealander Maurice Wilkins was Franklin's boss at King's College. When she hesitated publishing, Wilkins grew impatient and, without her knowing, showed her images to two physicists at Cambridge University, James Watson and Francis Crick. The duo had inklings that DNA was a helical structure, but it wasn't until they saw Franklin's images that they understood how DNA's components fitted together.

"The discovery of the double helical structure of DNA by Watson and Crick, using X-ray diffraction data from Rosalind Franklin, defined the nature of biological information and explained the chemical basis of heredity," says Collins.

It is said that, on 28 February 1953, Crick marched into the bar at the Eagle Pub in Cambridge and said just that: "We have found the secret of life." And it was true – up to a point. While Avery and

PROTEIN BUILDING

The DNA double helix is made up of sequences of just four molecules – adenine (A), cytosine (C), guanine (G) and thymine (T). These 'letters', known as 'bases', pair together in the combinations A–T and C–G along the double helix, and direct the manufacture of life-giving proteins from building blocks known as amino acids. There are two key steps:

(1) Transcription: Inside the cell nucleus, part of the DNA double helix unwinds, exposing the separate strands. Another single stranded molecule, messenger RNA (mRNA), forms along one of the strands, making an exact copy of the 'letters' in the gene sequence. The mRNA then moves outside the nucleus by travelling through the pores in the nuclear wall.

(2) Translation: Outside the cell nucleus, the ribosome provides a platform for protein synthesis. Molecules of transcription RNA (tRNA) are attached to amino acids – the 'building blocks' of life. When base pairs in a tRNA molecule bind with their corresponding mRNA base pairs, the tRNA releases its amino acid, which forms a chain with others, that then folds into a protein.

When a gene mutates, a stretch of DNA is either damaged or not read or copied exactly, and this error is expressed in a gene, and hence in the physiology of an organism.

his colleagues had identified DNA as the genetic material inside cells, Crick and Watson had shown how the genes were arranged within the molecule, as sequences of the building-blocks of DNA ('bases') strung along two intertwined helices.

"The connectivity of all living things was immediately given dramatic support, and the stage was set for the elucidation of the molecular basis of life and an ultimate transformation of medicine," says Collins. "I can think of no other moment in science that has had a larger impact."

Radiation Overdose

Watson and Crick published their work in the 25 April issue of the journal *Nature* in 1953. Franklin's work was also published in that issue. But it was Watson, Crick and Wilkins who in 1962 went on to win the Nobel Prize for their work on the structure of DNA, not Franklin. She had died four years earlier at the age of 37 – and Nobel Prizes are not awarded posthumously. Tragically, her work may have actually killed her. It is thought that the hours and hours spent exposed to X-rays took its toll. Franklin contracted ovarian cancer and died in 1958. In his book *The Double Helix*, published in 1968, Watson acknowledged the major role that Franklin played in the discovery of the structure of DNA.

GENE SEQUENCING

Revealing the DNA code that makes us human

Nominated by **Dr James Watson**,
co-discoverer of the structure of DNA

If you took all the DNA in one human cell and laid the chromosomes out end to end, they would reach nearly two metres.

DNA carries its all-important genetic information via sequences of four molecules, known as 'bases' – adenine (A), cytosine (C), guanine (G) and thymine (T) (see 236). In gene sequencing, the series of bases are read using chemical methods. There are two types of gene sequencing. In the Maxam–Gilbert method, enzymes are used to chop up the strand of DNA before each bit is treated with different

"When the structure of DNA was revealed in 1953, none of us would have predicted how quickly we'd reach the point where our individual genetic information could be obtained so fast. The pace of discovery has been astounding. Knowledge of DNA through sequencing means we'll be healthier. It's my hope that in five or ten years we'll be able to cure most major cancers."
James Watson

chemicals. As the four bases react to the various chemicals in different ways, each base can be identified and the sequence read.

English biochemist Frederick Sanger's work on the structure of the hormone insulin won him a Nobel Prize in 1958, and had a big influence on James Watson and Francis Crick when researching DNA. When Sanger himself turned his attention to DNA, he discovered another technique for sequencing genes, which earned him a second Nobel Prize in 1980. He's known to have said of this Nobel Prize: "You get a nice gold medal, which is in the bank. And you get a certificate, which is in the loft. I could put it on the wall, I suppose. I was lucky and happy to get it, but I'm more proud of the research I did."

In the Sanger method, a section of the DNA strand is cloned. During this process, synthesis is halted by adding a chemical that reacts with one of the four bases. By seeing which chemical stops the synthesis process, the base present at that point is revealed, and so the sequence of the whole DNA strand can be discovered.

The Human Genome Project

All the DNA base pairs in our cells – around 3 billion of them – are collectively known as the 'human genome'. Scattered within it are the relatively small number of base pairs making up the 20,000–25,000 genes which form the instructions for the production of proteins. The rest is sometimes dismissively known as 'junk DNA', which belies the fact that much of it is now recognised as playing a key role in the correct functioning of genes. The human genome isn't the largest known: that prize goes to *Paris japonica*, a flowering plant. Even so, 'decoding' the human genome's 3.2 billion base pairs is a huge challenge, and one taken on in the Human Genome Project.

The project was launched in 1990, but progress was initially slow due to the lengthy gene-sequencing process. In the mid-1990s, American Craig Venter began promoting a technique known as 'whole-genome shotgun sequencing' which turbo-charged Sanger's methods using computers to work out the sequence of base pairs of DNA from a huge number of random fragments.

Shotgun sequencing was controversial as some geneticists felt it wouldn't give a sufficiently accurate view of the complex human genome. But a rough draft of the human genome was produced in 2001. Just a couple of years later, on 14 April 2003, Craig Venter joined Human Genome Project leader Francis Collins at a press conference at the White House to announce that the full human genome had been sequenced.

Personalised Medicine

The medical benefits of knowing the sequence of our genes is likely to be considerable. For example, in 1995, the Sanger Institute in Cambridge located the BRCA2 gene, known to increase the risk of breast cancer. Meanwhile, researchers in Canada found that a person with all five variations of the FAD gene is almost 100 per cent certain to develop Alzheimer's.

One of the key findings of the Human Genome Project, though, was discovering that around 90 per cent of the genetic variation between each of us is down to so-called 'single-nucleotide polymorphisms' (SNPs). Some of these SNPs affect how susceptible we each are to different diseases – and drugs. So, while one person has a strong immune reaction to one drug, it might not have any effect on someone else.

The International HapMap Project has been set up to help scientists find more genes associated with specific diseases. Private companies have also jumped on the bandwagon, offering genetic tests for people – if they're brave enough – to find out if they carry genes linked to a higher risk of developing certain conditions. How predictive and useful these tests are is up for debate, though. Gene therapy is another burgeoning field. In some cases, viruses can be used to transport desirable genes to modify the genomes of specific cells.

It seems like an era of personalised medicine and genetically modified man awaits.

DID YOU KNOW?

Gene sequencing has found that humans have around 3.2 billion base pairs. The laboratory 'workhorse' of the bacteria world, *E. coli*, has 4.6 billion base pairs. But the organism with the most is the flowering plant *Paris japonica* with 150 billion.

GENETIC SWITCHES

How the environment and 'junk DNA'
play key roles in the genome

Nominated by **Mark Henderson**, science editor of *The Times*
and author of *50 Genetics Ideas You Really Need to Know*

Just 1.2 per cent of the human genome is made up of genes that contain instructions for making proteins. The remainder was once dismissed as being useless and nicknamed 'junk DNA', but it is becoming increasingly clear that this DNA is not junk at all.

"A lot of junk DNA appears to play a critical role in turning genes on and off in different tissues of the body and at different stages of development," says Henderson. "It contains genetic switches that are as important to our biology as the genes they control."

Indeed, recent research has revealed that almost all of our genome (about 90 per cent of it) may be being 'transcribed' into RNA. At least half of this RNA is already being shown to be useful. 'MicroRNAs', for example, regulate how genes are expressed by binding to over 60 per cent of protein-coding genes. MicroRNA and other non-coding RNAs are produced at different stages in different types of cells. As most cells have the same genome but appear very different, this suggests that non-coding RNAs play a crucial role in regulating genes and hence determining the nature of cells. Moreover, non-coding RNAs may be the answer to another puzzling phenomenon.

We each have between 20,000 and 25,000 genes. That sounds like a substantial number, until you find out that the lowly 1 mm-long nematode worm with its 1,000 cells has almost the same number of genes. It seems likely that non-coding RNAs are responsible for making humans more complex, despite having roughly the same number of genes.

Epigenetics

We now know that it's not just our genes that determine our physiology. Our bodies and brains are also the result of a complex conversation going on between our genes and the environment we're exposed to. This is known as 'epigenetics'.

"Epigenetics involves another type of genetic switch, whereby genes and chromosomes become chemically modified by environmental triggers," says Henderson. Indeed, epigenetics adds another layer of information to the genome by adding chemical marks to DNA, in a process known as 'methylation', thereby changing its sequence. If you imagine the genome to be the book of life, then these chemical marks are a bit like underlining key paragraphs. Another analogy is thinking of the genome as the hardware of a computer, and the so-called 'epigenome' as the software that instructs the computer how to work.

This phenomenon takes effect very early on in development, even before a woman knows that she's pregnant. Research has shown that your ancestors' lifestyle affects your epigenome. A study on the inhabitants of the town of Överkalix in Sweden found that grandfathers who overate had grandchildren who were four times more likely to die from diabetes.

"We are slowing beginning to understand how epigenetics choreographs the intricate dance between nature and nurture that shapes life and health," says Henderson.

Indeed, it now seems possible to reverse the methylation process and hence control disease. For example, the *BRCA1* gene is known as a 'tumour suppressor gene', as it stops tumours developing in breast and ovarian tissue. Adding chemical marks switches off the gene, preventing it from producing the tumour-suppressing proteins. Epigenetics could therefore be used to predict who is at risk of developing cancer, so that they can be treated with drugs that strip the gene of tumour-causing chemical marks before they develop cancer.

DID YOU KNOW?

- One hundred mutations in your genome make your DNA different to that of your parents.

- Chimps and humans differ in directly comparable DNA sequences by 1.2%.

- The genomes of two unrelated humans are only 99.1% identical.

The Medical World

GERM THEORY

How a host of scientists revealed a deadly microscopic world

Nominated by **Carl Zimmer**, author of award-winning blog **The Loom** and books including **Microcosm: E. coli and the New Science of Life**

Each of us is made up of between 10 and 100 trillion cells. But there are 1,000 times more bacteria than that both inside and on each of our bodies. For this reason, both good and bad bacteria determine our health.

Bacteria are essential for life as we know it, because they feed on dead material, breaking it down into chemical components, which can then be reused – for example,

"Over the past 120 years or so we've seen a huge number of medical advances, such as vaccines and antibiotics, along with a spectacular drop in deaths from infectious diseases. While all these triumphs are worth celebrating individually, they all come from the same fundamental insight: the germ theory of disease. Put simply, this says that specific pathogens could cause specific diseases. From that insight came the many defences we now take for granted, including the simple act of washing our hands."

Carl Zimmer

leftover vegetables are decomposed into compost that can be used as fertiliser. Industry spotted bacteria's potential years ago, and now makes use of its talents for, among other things, the production of drugs and the disposal of sewage. The flipside is that bacteria can also be deadly to humans.

Man has found ways of controlling most things, but disease is something that continues to plague us. For centuries, people believed that diseases were caused by 'bad vapours' in the air. But an English doctor had another idea.

Dr William Budd was born in the town of North Tawton in Devon in 1811. While working as a surgeon during an outbreak of typhoid in 1839, he noticed that the disease spread from victims to the people looking after them. For the next eight years, he studied the contagious disease and came to the conclusion that it could be spread through water. He thought that if 'poisons' could multiply in the intestines of sick people and then make their way into sewage, they would eventually reach healthy people through contaminated water supplies. After moving to Bristol, he made efforts to protect the town's drinking water and was responsible for a huge reduction in deaths from cholera infection.

At around the same time, Dr John Snow succeeded in convincing the council to remove the handle from a water pump in Soho's Broad Street in London. He'd realised that the water was contaminated with sewage, which must be carrying the agent of cholera. When the pump handle was removed in 1854, the cholera epidemic in Soho subsided.

On Vacation in a Mental Institute

A Hungarian physician by the name of Ignaz Semmelweis had found similar results in the mid-1800s while working in a maternity clinic in Vienna. A fatal condition, called 'childbed fever', was raging through the clinic, but the number of deaths was higher in the doctors' than in the midwives' unit. Semmelweis and his fellow doctors tried everything, from feeding the patients differently to dissecting the dead bodies. But it was the death of one of the doctors that opened Semmelweis's eyes to what was going on.

After pricking his finger with a knife while examining a victim, the doctor in question had contracted a disease with symptoms suspiciously like childbed fever. Semmelweis realised that, whatever the killer was, it must be being carried on their hands. He insisted that everyone wash their hands with soap – and the death rate dropped dramatically.

But the fanatic in Semmelweis was eventually his downfall. As he worked his way around different hospitals in Europe, he encouraged colleagues to wash their hands between procedures and became so obsessed by this cleanliness that he even wrote to leading European doctors accusing them of committing murder if they didn't wash their hands.

The medical community – and his wife – had had enough. In 1865, she organised a 'holiday' in Austria, visiting a friend's 'hospital'. In fact, this was his last trip to his final resting place: a mental institute. He contracted childbed fever soon after arriving and died a lonely death. Although Semmelweis's work was in vain, a Frenchman was about to discover the invisible world hidden on dirty hands.

Spoilt Milk

Louis Pasteur was born in 1822 in Dole in eastern France. As a Frenchman, what else would one think of using as a good lab test sample than wine. Now Pasteur wasn't that keen on wine, but in nineteenth-century France you'd make a killing if you could work out why wine goes off. So the chemist set about researching what caused the wine to spoil.

His hours in the lab finally paid off when he discovered that it was the growth of microorganisms that spoiled the wine. And it wasn't just wine that this happened to, but also beer and milk. As Pasteur tinkered around with the liquids, he realised that heating them killed the microorganisms, a process now called 'pasteurization'. If the liquids weren't exposed to air, and hence to microorganisms, they didn't go off.

In the 1870s, Pasteur realised that microorganisms must also be responsible for infectious disease – the essence of germ theory – and applied his research in this direction. Working from Edward Jenner's idea of vaccination (see page 255), he developed vaccines for cholera, anthrax and rabies. It was this last disease that could have seen this gutsy Frenchman prosecuted if his tests had gone wrong.

Pasteur had developed the rabies vaccine in rabbits, before making it less virulent by drying out the infected nerve tissue. It was all very well testing it on animals, but Pasteur knew the vaccine wouldn't be adopted by the medical community unless he tested it on a person. Throwing caution to the wind, in July 1885, Pasteur decided to inject a young boy that had been mauled by a rabid dog.

It worked. The boy didn't develop rabies, and Pasteur secured his name in the annals of science history.

VACCINATION

How a village doctor saved humanity from a deadly disease

Nominated by Sir Richard Branson, entrepreneur and head of the Virgin Group

"Don't think, try." This was the mantra that drove surgeon John Hunter, and one that he instilled in all the trainee doctors he mentored. But his most talented young protégé, Edward Jenner, really took the motto to heart and, in the process, made a groundbreaking discovery that radically changed the face of human health.

For centuries the world had been gripped by a terrible disease: smallpox. Victims suffered from rashes that developed into small, raised blisters filled with fluid. But the effects of the disease were by no means small. During the late eighteenth century, almost half a million people in Europe alone died from the disease every year, including five reigning monarchs, and one-third of all blindness was linked to smallpox.

Throughout the seventeeth century, Europeans had adopted a treatment of inoculation that had been used in Asia for centuries. The skin of a smallpox victim was scratched open and the wound rubbed with the material from a scab of a victim who had a less virulent form of

CAUTIONARY TALES

In 1874 a young doctor by the name of William Osler was treating one of his first patients in a hospital in Montreal. The patient had what the French-Canadians were calling *la picotte noir*, the deadly black smallpox. Despite Jenner's breakthrough – vaccination having been adopted worldwide – doctors could do little for victims already suffering from the disease. So the key was to get the city of Montreal vaccinated and all victims quarantined. But it wasn't that simple.

In the 1870s, smallpox was so virulent in Montreal that a whole hospital building had been allocated for smallpox victims. Over the years, as the epidemic seemed to wane, the hospital closed down. But a few years later, the disease reared its ugly head again.

When train conductor George Longley pulled into Montreal station, on 28 February 1885, he wasn't feeling quite right. As his condition worsened, doctors isolated him in Hotel-Dieu Hospital. But that didn't contain the disease. By April the health authority was taking extreme measures, sealing off and disinfecting the hospital and other buildings with sulphur. Extreme cases that couldn't make it to isolation hospitals were quarantined in their own homes, which were labelled as housing smallpox sufferers. But the virus was still on the loose in Montreal.

the disease. It sometimes worked, inducing a mild case of smallpox and subsequent immunity, but all too often it would develop into full-blown smallpox, igniting an epidemic. Millions continued to die of the disease.

Victims weren't cooperating with the health authority, tearing down the placards on their houses and refusing to be isolated. Vaccination was still deemed by many, including eminent doctors, to be wrong. They claimed that injecting an animal disease into a human was unnatural.

As the epidemic spread rapidly, riots broke out and the death toll rose. It took over a year to gain control over the disease. By that time about 2 per cent of the city had died from smallpox, the majority of whom were children. The key problem had been that many parents had refused to get their children vaccinated.

Even with today's education, people are wary of vaccination, but refusing to be vaccinated can be costly. In February 1998, Andrew Wakefield published a study in the medical journal *The Lancet* which claimed a link between the MMR vaccine and autism. However, the study was based on only 12 people and data was found to have been tweaked. Eventually the link was proven false, but by that time cases of measles had risen after vaccination rates had dropped from 92 to 80 per cent, and two children had died.

In some developing countries, vaccination is still seen by some as interfering with religious beliefs or being a conspiracy to infect ethnic minorities. But there is also still an anti-vaccination movement closer to home. As Branson says: "At a time when faith in vaccines is threatened by an anti-vaccination lobby in the US and UK, and rates of many diseases are rising, we need to trumpet this simple but crucial breakthrough in human health."

When Edward Jenner moved out of London and set up his own practice in the village of Berkeley in rural Gloucestershire in 1773, he was determined to find a cure for smallpox. The more patients he saw, the more a pattern began to form: many of the milkmaids who visited his practice weren't falling prey to smallpox. Further investigation revealed that all of them had at one time or another contracted cowpox, a benign form of the disease. It dawned on him that if cowpox could be transferred from one person to another then the benign form might prevent the deadly type developing.

Practising what his mentor had preached, Jenner decided to try out his idea. But for that he needed some volunteers. One of his patients was a milkmaid, Sarah Nelmes, who was suffering from cowpox she had contracted while working with a cow not-so-aptly-named Blossom. On 14 May 1796, Jenner collected fluid from the milkmaid's blisters and transferred it into James Phipps, the son of his gardener. Then a couple of months later, on 1 July, he deliberately infected the boy with smallpox.

It was a risky thing to do, but the boy lived to tell the tale. And what a tale it was – he was the recipient of the world's first vaccination, so-called because *vacca* in Latin means 'cow'. The risk had been enormous, but Jenner had weighed up the pros and cons, and realised that the potential gain was immense.

"When Edward Jenner started treating people with the smallpox vaccine, in one health measure he eradicated the cause of death of nearly a quarter of the world's population," says Branson.

In 1798, Jenner produced a pamphlet about the benefits of vaccination. But it took over 40 years for him to persuade the government that vaccination was the way forward. In 1840, politicians realised the significance of the doctor's successful localised trials, and two years later variolation was banned and vaccination became mandatory.

Jenner never knew the exact mechanism by which his vaccination worked, but he did realise how momentous his discovery was: "It is a privilege to make a real difference in your own lifetime."

In 1980, the World Health Organization announced that smallpox had been completely eradicated. Some samples of the disease do still exist, but they are locked away in secure labs in the US and Russia. To date, billions of lives globally have been saved through vaccination against all sorts of diseases.

As Branson points out: "Over generations, vaccination has been the biggest health advance."

ANTIBIOTICS

The chance discovery of penicillin that saved millions of lives

Nominated by **Robin Ince**, writer and comedian, co-presenter of the science tour *Uncaged Monkeys*

During World War I more soldiers died of infection than from fighting on the battlefield. Disease-causing bacteria were to blame. Antiseptics had previously been used to kill bacteria, but they also killed human cells. A new remedy was desperately needed. Fortunately, before the end of the 1920s, it appeared, and it was discovered completely by accident.

Alexander Fleming had been away from his laboratory at St Mary's Hospital in London for a few days. The Scottish bacteriologist had left out some petri dishes of staphylococci, the bacteria he'd been studying. On his return to the laboratory on 28 September 1928, he started clearing away the old petri dishes, when suddenly something caught his eye.

In one of the petri dishes, a dark mould had grown on the gel used for developing bacteria. Crucially, Fleming noticed the area around the mould was completely free from bacteria, which could only mean that something in the mould was killing them. After a number of tests, Fleming confirmed that the mould he'd identified as *Penicillium notatum* was secreting some previously unknown substance with antibiotic properties.

Fleming called this substance 'penicillin' and we now know that it works by breaking down bacterial cell walls. Although Fleming's discovery was an important milestone in the history of medicine, at the time penicillin could only be produced in minute quantities, and Fleming himself did not believe it could have any clinical application. However, all this changed when, a decade later, an Australian pharmacologist and a German biochemist, working at Oxford University, came across Fleming's work.

In the late 1930s, Howard Florey and Ernst Chain started to look into how they could make penicillin in sufficient quantities to be used in the medical treatment of humans. Together with the British biochemist Norman Heatley, they produced enough penicillin to demonstrate that it protected mice injected with lethal doses of bacteria. Under the severe privations of wartime Britain, Heatley's inventiveness in extracting and purifying penicillin (using, for example, bedpans from the Radcliffe Infirmary to grow the mould artificially) created enough of the drug to begin the first clinical trials on humans in 1941.

Albert Alexander, an Oxford policeman, who had acquired life-threatening septicaemia after being scratched by a mere rose thorn, was their first patient. He was injected with penicillin and started on the road to recovery. On the fifth day he was sitting up in bed, but then the antibiotic began to run out. Ingeniously, the team extracted the precious drug from his urine and readministered it. When supplies ran out completely, the policeman's condition deteriorated and he died. Nevertheless, it was palpably clear that the drug was efficacious.

The Oxford team then shared its knowledge with American pharmaceutical companies, enabling them to mass-produce the drug in time for the final stages of the war, saving thousands of lives. Fleming, Florey and Chain shared the 1945 Nobel Prize in Physiology and Medicine for "the discovery of penicillin and its curative effect in various infectious diseases". The stone memorial outside the University Botanic Garden in Oxford, commemorating the work of the penicillin team, concludes with the words: "All mankind is in their debt."

"When confronted by a noisy anti-science lobbyist, I think of many arguments against their wrong-headedness. The most stinging is the steep decline in youthful deaths over the last hundred years, due largely to the discovery of penicillin," says Ince. "The only reason they have managed to grow to wrong-headed adulthood rather than get cold in an early grave is the science they are arguing against."

Other antibiotics, such as streptomycin and the tetracyclines, were developed in the years that followed. However, microorganisms reproduce rapidly, so bacteria can evolve relatively quickly to become resistant to antibiotics. Doctors are consequently more and more reluctant to prescribe antibiotics for minor infections, while the search for new drugs continues.

CONTRACEPTION

From poison to condoms to the pill:
how women were freed to spread memes,
not just genes

Nominated by **Dr Susan Blackmore**, visiting professor at
Plymouth University and author of **The Meme Machine**

Rashes, puss-laden lesions, organ failure and eventual death. These
symptoms of syphilis are enough to put any man off sex – or at least
make him put on a condom.

Condoms have been around for a long, long time, but they haven't
always been made of latex or come with a 99 per cent assurance tag. In
the seventeenth century, they were made from animal intestines. Even
Italian adventurer, author and infamous womaniser Giacomo Casanova
was known to use these 'assurance caps' during the eighteenth century,
to prevent impregnating his numerous mistresses.

Nevertheless, as with so many other inventions, it was probably the
Egyptians who were the first to use contraception. Records exist of
women using a pessary: essentially a small object inserted into the vagina,
and covered with an acidic substance, such as honey or oil, to kill sperm.
The Egyptians didn't know exactly how this prevented fertilisation, as

the function of sperm wasn't discovered until the seventeenth century, when Antonie van Leeuwenhoek developed a sufficiently powerful microscope to allow him to study sperm structure.

Herbal concoctions that induce abortion have been around for a lot longer. The Arab world was streets ahead of Europe in this area: eight centuries, in fact. In the encyclopedic *Canon of Medicine*, written by Avicenna in 1025, he lists 20 birth control substances. Other less salubrious liquids have also been ingested over the years. In the second century AD, Greek gynaecologist Soranus suggested mothers drink water used by blacksmiths, while desperate women burdened with unwanted pregnancies have been known to take poisonous mixtures of mercury and arsenic.

Infanticide

If efforts to get rid of unwanted pregnancies fail, some people resort to murder of the child once it is born. Horrific as this sounds to the rich world, it is a dark reality in some parts of the world. In India, sons tend to hold higher social and economic value for their parents than daughters, while in another burgeoning superpower, China, the one-child policy has been blamed for high levels of infanticide skewing the sex ratio.

Infanticide has existed throughout history, initially as child sacrifice for placating the gods, more recently as a result of poverty or politics. The practice shows the extreme measures to which people are prepared to go in order to get rid of unwanted offspring.

"Contraception must surely be the greatest invention of all time," says Blackmore. "Modern contraceptives have freed up half the population from spending all their best years bearing and bringing up children."

Blackmore herself has contraception to thank for a successful career and varied life. "It allowed me to have a scientific career, write books, be in the public eye and travel the world. All that wouldn't have been even remotely possible if I hadn't had birth control. When my two children were young, I spent a lot of time with them. But once they went to school, I was able to get on with my work, knowing with confidence that I wasn't going to get pregnant again."

It Was a Man's World

"Over 100 years ago, some women had brilliant scientific ideas or great discoveries," says Blackmore. "Take the case of Mary Anning, whose fossil discoveries in Dorset in the nineteenth century dramatically helped to advance theories on prehistoric life. She wasn't allowed to read her own paper at the Royal Society: a man had to read it for her."

Sexual discrimination continued for the rest of that century. Then, as the suffragette movement gained momentum, the vocal Margaret Sanger in 1914 swept aside Victorian protocol with her monthly newsletter, *The Woman Rebel*, in which she encouraged the use of contraception, coining the term 'birth control'.

But it wasn't until the swinging sixties, famous for rock and roll, drugs and free love, that mass production of the pill liberated thousands of women in the West from unwanted pregnancies. "Once women could decide for themselves whether they wanted to have children, they could devote themselves to art, science, music, or any career," says Blackmore. "And rather than just spread their genes, they were able to spread memes."

Infertile Future

So the developed world is in control of its uterus. But what of the developing world? Overpopulation must surely be the biggest challenge facing humanity. It creates a vicious circle that's difficult to

THE MEME MACHINE

A 'meme' is information, such as an idea, skill, technology or story, that is copied or imitated from person to person or through other media. "*Little Red Riding Hood* is a very successful meme that has spread through many different cultures, as is the Theory of Relativity," says Blackmore. "They've spread for different reasons – one appeals to people's love of storytelling, the other appeals to the desire to discover the truth. But, of course, false theories such as homeopathy can spread as memes, not because it is true but because people want to it to be true, which causes the placebo effect. Memes spread selfishly, in the same way that genes do."

The term 'meme' was coined by evolutionary biologist Richard Dawkins in 1976 in his book *The Selfish Gene*. Blackmore has written whole books on the subject herself. "Meme-spreaders are the people whom everyone else listens to, from pop stars to scientists. The best meme-spreaders in science are those who do really good work that others want to know about, and the ones who are vociferous and confident enough to write a good paper and give lectures. That always used to be men, but more and more women are becoming meme-spreaders."

break. With the global population now over 7 billion and expected to reach 9 billion by 2045, water will become the next oil and food evermore scarce for the millions that live on a dollar a day. As we pump more greenhouse gases into the atmosphere, global warming will melt the ice caps and raise sea levels, squeezing civilisation inland to less fertile ground.

There are still far fewer women than men in the world of science, though. "Some of the discrepancy is down to biological fact – men and women are different, their brains are different, the way they think is different, the things they like to do are different," says Blackmore. "There are exceptions, of course, but fewer women are inspired with the kind of curiosity that you need to be a good scientist, let alone a great one."

Another factor is that a lot of women scientists want to work part-time for a balanced life. Scientific institutions aren't always conducive to this. "It's a highly competitive, male-dominated system. Many women don't want to fight the battles that come with climbing up the ladder. Our society as a whole is bad at finding ways to incorporate sex differences."

The meme race is speeding up. We're bombarded with more and more information which is transferred at ever-faster rates. Science itself is growing more competitive. "You can be a successful scientist working part-time and taking some years out," says Blackmore, "but the more competitive the field you're in, the harder that is."

Many developed societies are actually decreasing in population. While developing countries must never be denied luxuries such as running water and electricity at the flick of a switch, contraception could be the key to keeping the population around the world at a sustainable level.

SYSTEMATIC REVIEW OF CLINICAL TRIALS

The evolution of truly unbiased results

Nominated by **Dr Ben Goldacre**, medical doctor and author of **Bad Science**

"By picking out only the positive trials, and ignoring the negative ones, you can make an ineffective drug look like it works well. When the idea of a systematic review of clinical evidence was introduced, it put a stop to all this."

Ben Goldacre

Antidepressants are among the most widely prescribed drugs in the world, with tens of millions of people taking them each year. Many people hope the drugs will help them combat their crippling affliction, and doctors have been able to point to proof of the effectiveness of drugs like Prozac in rigorous scientific clinical trials. Published in respected medical journals, the results of these trials have shown that around 60 per cent of people taking the drugs benefit from them, compared to just 40 per cent or so of those taking useless placebos – a clear level of benefit.

Or at least that's what they seem to show. The problem is that these trials only tell part of the story. In 2008, researchers independent of the drugs companies found that many other trials into the effectiveness of these antidepressants had been carried out, but never published. And when the results from these were combined with the publicly available evidence, the overall effectiveness of the drugs turned out to be only modestly better than not taking anything at all.

Digging further, the researchers found something else: of all the trials they could track down, 94 per cent of those finding a positive result ended up being published, compared with just 14 per cent of those with negative or unclear outcomes. Known as 'publication bias', this tendency for good news to get published and the rest to stay in filing cabinets has been recognised in many scientific fields. Various explanations for it have been put forward, ranging from journal editors being keener on headline-grabbing good-news stories to researchers not bothering to write up disappointing results. But there's no doubt that publication bias holds special dangers in medicine, as it can lead to useless or even dangerous drugs becoming widely used. In the case of antidepressants, there's little doubt that the drugs do offer some benefit, but, equally, there's now little doubt that patients are more likely to be disappointed by their effectiveness than the published results suggested.

To guard against such cases in future, medical researchers have developed techniques collectively known as 'systematic reviews'. These involve tracking down and analysing all studies, including those stuck in the filing cabinets of researchers and regulators. The most important of these studies are those based around one of the greatest medical breakthroughs: the randomized controlled trial, or RCT.

Scurvy Cure

The idea of a clinical trial can be traced back centuries. In 1747, James Lind, a Scottish physician who pioneered naval hygiene, showed how citrus fruits could be eaten to cure sailors of scurvy. Symptoms of the disease include spots on the skin, bleeding gums and, in extreme cases, jaundice, fever and death. It's now known to be caused by a deficiency in vitamin C. In Lind's trial, all scurvy patients ate the same diet, but some were allowed to supplement it with cider, vinegar, nutmeg and also, crucially, oranges and lemons. The ones who ate citrus fruits recovered in six days. This was the first time control groups had been used. However, scepticism about the validity of Lind's approach, and its reliance on small numbers of patients, meant that nearly half a century passed before the navy made lemon juice a compulsory part of sailors' diets.

Not until the mid-twentieth century did clinical trials go mainstream. In the 1940s, the UK Medical Research Council (MRC) set up trials involving two of the central pillars of clinical trials: 'double-blinding' and the randomised selection of patients. In the study of a putative cold cure known as 'patulin', the MRC researchers ensured that neither the patients nor the doctors treating them knew who was getting which treatment. This so-called 'double-blinding' is designed to prevent patients or their doctors fooling themselves into seeing effects that aren't real. And it proved a vital counter to any optimism on the part of the doctors – as patulin turned out to be useless.

Then, in 1946, the MRC began trials of a whooping cough vaccine and the use of the antibiotic streptomycin for treating tuberculosis. Both trials were double-blinded, but following recommendations by the pioneering British medical statistician Austin Bradford Hill, another key feature was introduced: random allocation of patients to either the group receiving the therapy or the comparison group. This randomisation minimised the risk that patients getting the new therapy were unusual in some way, and thus likely to give a false impression of the drug's effectiveness.

Systematic Review

The aim of an RCT is to give completely unbiased insights into the effectiveness of a new therapy. But RCTs are very expensive to set up and rarely produce definitive results by themselves. It usually takes the combined results of many RCTs to produce a clear picture. Yet, if only some of the RCTs are published, the result can be biased – and highly misleading.

The concept of the systematic review started to gain popularity in the early 1990s and is now enshrined in the work of organisations such as the Cochrane Collaboration. Independent researchers review all the available evidence to give the most up-to-date and unbiased assessment of therapies, which can be relied on by doctors anywhere in the world. The result has saved countless lives through the weeding out of so-called 'best practices' that systematic reviews revealed were either useless or murderously flawed.

CHEMIOSMOSIS

How cells generate energy

Nominated by **Dr Nick Lane**, research fellow at University College London and winner of the 2010 Royal Society Book Prize for his book *Life Ascending: The 10 Great Inventions of Evolution*

Sometimes it pays to stick to your guns. This was the case for the biochemist Peter Mitchell, who discovered the process of chemiosmosis. Before the 1960s, adenosine triphosphate (ATP) was known as the energy currency of life, but no-one quite knew how it was made. Then, in 1961, Mitchell proposed a new theory.

> *"As universal as the genetic code, chemiosmosis is the most counterintuitive idea in biology since Darwin. Who would have guessed that life generates power by turning chemistry into electricity and back again?"*
> Nick Lane

Mitchell claimed that ATP is formed when hydrogen ions pass from an area of high concentration on one side of a membrane to an area with a lower concentration, through a protein in the cell membrane called 'ATP synthase'. This works a bit like a hydroelectric dam in which the flow of water (hydrogen ions) drives a turbine (ATP synthase).

We now know that this chemiosmotic process occurs in plant cell chloroplasts during photosynthesis. In animals, chemiosmosis takes place in the powerhouses of the cell – the mitochondria – during cellular respiration, where glucose is converted to carbon dioxide and water. It is a vital process for any cell, and produces a lot of energy: "Across tiny distances, it generates a voltage equivalent to a bolt of lightning," says Lane.

When Mitchell proposed his theory of chemiosmosis, there was not a shred of evidence to back him up, and his theory generated a huge amount of controversy within the scientific community. But, after years of acrimony and detailed experimentation by Mitchell and others, his hypothesis was finally accepted: life really is powered by protons. In 1978, Mitchell was awarded a Nobel Prize in chemistry.

THE BRAIN

Discovering that the mind is generated from the brain

Nominated by **Professor Steven Rose**, emeritus professor at the Open University and Gresham College London, visiting professor at University College London, and author of many books, including the Royal Society Book Prize winner *The Making of Memory*

For many centuries, few people realised the importance of the 1.5 kg jelly-like grey and white matter that sits on top of our spinal cord. The ancient Egyptians knew that the brain was covered by membranes, and that brain trauma affects the human body, rendering limbs immobile. Egyptian physicians had even devised ways to treat brain injuries – some successful, some doing more damage than good. But they still didn't recognise how vital the brain was, believing that the heart was the key to consciousness – preserving the organ in a separate jar while the brain was simply thrown out.

Greek Herophilus from Alexandria did clock how important the brain was, though. Around 300 BC, he performed numerous human dissections – slicing up corpses and, shockingly, live convicts, noting how tinkering with the brain affected different parts of the body. But, as Christianity replaced the Roman civilisation, the ideas of the Church ruled over surgical investigation.

Many centuries later, during the cultural revolution of the Renaissance in the 1600s, the renowned French philosopher René Descartes recognised the brain's key role in all human functions. Legend has it that while walking in a park near his Parisian home, his eye was caught by some mechanical sculptures. As he watched the levers moving, an idea sparked in his mind – just as the hydraulic tubes brought the sculptures to life, maybe our brains controlled our limbs.

"Western religion and philosophy traditionally divided mind from matter, body from soul. The mind directed the body through the brain. Modern neuroscience has reversed this relationship, claiming that the mind emerges from brain processes. This revolutionary materialist reversal – whether right or wrong – underpins scientific thinking about human behaviour, mental illness, and how consciousness evolved."
Steven Rose

This blasphemous idea was a little too much for the Church, though, and Descartes spent the rest of his life on the run from the infamous Inquisition. Some believe the dreaded group finally caught up with him in Stockholm where he died in 1650 of 'pneumonia' – according to the official version of events.

But Descartes's theory survived. His idea that the brain controls the body, making us who we are, was encapsulated in his well-known phrase "I think therefore I am", influencing the work of many scientists and physicians.

Pulling Apart the Brain

Thomas Willis, the seventeenth-century Oxford physician and natural philosopher, is best remembered today for his work on the brain. He methodically dissected the brains of all sorts of creatures from earthworms to sheep. Willis also became the first person to investigate every part of the human brain, asking his Royal Society colleague Christopher Wren to record all the parts in detailed drawings. Willis's *Cerebri Anatome*, published in 1664, was the first work to use the words 'neurology', 'lobe' and 'hemisphere', and to recognise that the 'lower brain' was common to all creatures, while the 'higher brain' was only fully developed in humans. It was the first time that anyone had shown evidence of the fact that we are who we are because of our complex brains. Crucially, though, Willis had also been the first to assign different functions to different brain regions. Indeed, the 'Circle of Willis', a ring of arteries at the base of the brain, is named after him.

In the mid-eighteenth century, Swedish scientist Emanuel Swedenborg proved that different sections of the brain's cortex control different muscle groups. Then, at the turn of the nineteenth century, a colourful German-born anatomist developed this idea of 'localisation'. During his extensive travels with a retinue of women and pets, Franz Gall looked into the idea that different brain regions are responsible for different personality traits. After studying hundreds of patients with known mental conditions and recording the shapes of their heads, he claimed he could diagnose a mental condition simply by examining their skull. As a consequence of Gall's theories, many people were wrongly labelled as having a psychiatric condition because their head shape didn't quite fit the norm.

While Gall's theories ruined the lives of many, his methods were never invasive. From the late 1800s, however, other scientists were literally pulling apart the brain, performing horrendous animal experiments and human lobotomies.

Lab Rats

It wasn't just rats that were subjected to live investigation. The brains of rabbits, dogs and even tortoises were split apart, poked with scalpels and exposed to electric shocks and chemicals, all in the name of science. German physiologist Friedrich Goltz removed large pieces of a dog's brain and kept the animal alive for 18 months, even showcasing how the poor creature could continue to function at the International Medical Congress in 1881. At the same conference, Scottish neurologist David Ferrier showed how a macaque monkey had been paralysed after he'd removed part of its cortex. The medical community were impressed and agreed with his theory that intellect wasn't just linked to one specific brain region.

Anti-vivisectionists protested loudly, but cleaving brain tissue wasn't reserved solely for animals. Lobotomies became the order of the day for patients with mental conditions after scientists heard of the case of construction worker Phineas Gage. When some dynamite exploded in his face one day while working on a section of railroad in Vermont, Gage was left with a metal rod lodged in his skull and a completely changed personality. The cause of this character switch was blamed on the severe damage to his brain's frontal lobe.

This case and ones like it intrigued neurologist Walter Freeman. He developed the 'transorbital lobotomy', where an ice pick was inserted through a tear duct to cut into the brain of a patient with a mental condition. From the 1930s to the 1960s, Freeman performed over 3,000 such lobotomies, but few of the patients showed improvement, many deteriorating further.

So a century of horrific experimentation had allowed scientists to understand aspects of the brain and its different functions. Nonetheless, many of the intricacies of our grey matter were still unclear. It fell to an Italian and a Spanish scientist to reveal more about the actual structure of the brain (see page 279).

NEURON THEORY

Revealing the central nervous system

Nominated by **Professor Uta Frith**, professor of cognitive neuroscience at University College London

Imprisoned at the age of 11 for destroying the town gate with his home-made cannon, Santiago Ramón y Cajal was an interesting character. Growing up in the Navarre region of Spain, he loved drawing and dreamt of becoming an artist. His father, a professor of anatomy, had other ideas, forcing him to study medicine. Despite his rebellious character, Cajal toed the line and followed an academic medical career.

Cajal ended up with the best of both worlds, as his academic work of researching disease, microbiology and, crucially, anatomy needed him to draw detailed images. His draftmanship was of an incredibly high standard, but his contribution to science was even more impressive. In his work on the central nervous system, he discovered various different parts of nerves and, around the turn of the twentieth century, found conclusive evidence for what is now called 'neuron theory'.

A number of years earlier, in 1873, Italian scientist Camillo Golgi had made a groundbreaking discovery. While working in a psychiatric hospital, Golgi realised that in order to see what was what in the nervous system, he needed a way to reveal certain nerves but not others. Staining tissue with silver nitrate proved to be the way to do this, as it only stained a limited number of the cells, so he could see certain pathways of neurons in brain tissue. In honour of this breakthrough, a key part of the cell was named after him: the 'Golgi apparatus'.

Around 1900, using Golgi's method, Cajal was able to see that nerve tissue was made up of separate individual cells – nerve cells. Cajal's discovery started a bitter dispute between him and Golgi, since Golgi believed the brain to be a network of connected tissue, while Cajal defended the idea of separate neurons. Despite an enormous public row, they both got the Nobel Prize in 1906.

DID YOU KNOW?

There are about 1,000 billion neurons in the human body – more than the number of stars in the Milky Way!

"The idea that the brain and central nervous system were made up of discrete neurons was revolutionary," says Frith. "Another major breakthrough was the idea that what neurons transmit to each other cannot be understood in terms of chemistry and energy only, but must also include information. Once it was possible to think of the brain as an organ that processes information, it became possible to work on theories about how we see things in the world, and remember them."

THE X-RAY MACHINE

How the discovery of mysterious rays
unveiled the structure of the human skeleton

Nominated by **Dr Tim Boon**, chief curator
at the Science Museum, London

*"X-rays marked the culmination
of a revolution in medicine.
Earlier diagnostic devices
had been small and delicate
instruments, but the X-ray
machine ushered in the age of
machine medicine that we all
experience today."*

Tim Boon

Eureka moments are so often
complete accidents, and this was the
case when X-rays were discovered
completely by chance.

Michael Faraday had sprung to
fame in the 1820s when he created a
device that made an electric current
rotate a magnet: a rudimentary
electric motor. Over a decade later, Faraday made another important
breakthrough. By fitting metal electrodes inside a sealed glass tube that
had been evacuated and then applying a high-voltage to the electrodes,
he managed to produce an eerie glow inside the tube.

Excited by Faraday's results, other scientists started investigating this
phenomenon. They found that, even when the glass tube was almost

completely empty of air, one end of it still glowed. Some unknown rays must be travelling along the tube, which the German physicist Eugen Goldstein named 'cathode rays'.

It was while investigating these mysterious cathode rays that, in 1895, another German physicist, Wilhelm Conrad Röntgen, stumbled across a different type of ray, one that was to dramatically affect the world of science and win him a Nobel Prize in 1901. To block out natural light that might interfere with his experiment, Röntgen had put up a chunky piece of cardboard. As he switched on the electricity and his vacuum tube began to glow, he noticed that a nearby screen was glowing as well, despite the cardboard barrier. The cathode rays couldn't reach it because they were angled in one direction only. It dawned on him that some other ray must be causing it, bizarrely passing through the cardboard. Puzzled, he called them 'X-rays', the 'X' standing for 'unknown'.

Metal stopped the mysterious rays, but paper and human skin didn't. Placing his wife's hands between a beam and some photographic film, Röntgen was able to get an image of the bones inside her hands.

"A diagnostic revolution had been under way through the nineteenth century, symbolised by the invention of the stethoscope in 1816," says Boon. "But with the discovery of X-rays, doctors could really see inside the body for the first time."

Hearing of Röntgen's incredible discovery, 15-year-old Russell Reynolds immediately started designing his own X-ray machine. With the help of his father, who was a medical doctor, the schoolboy set about building it and, within a year, in 1896, they'd finished.

"Reynolds's machine is on display in the Science Museum," says Boon. "In 2009, it was voted as the favourite by visitors from a shortlist of 10 Science Museum objects all representing milestones in the history of science and technology."

Gradually, as their medical potential became clear, a whole range of X-ray machines were built over the couple of decades following Reynolds's invention. It wasn't until the turn of the twentieth century that the deadly nature of X-rays was recognised. Too much exposure damaged skin cells and resulted in radiation burns. So smaller and less frequent doses were used with more caution. Later that century people finally woke up to the true horror of overexposure and strict guidelines were introduced.

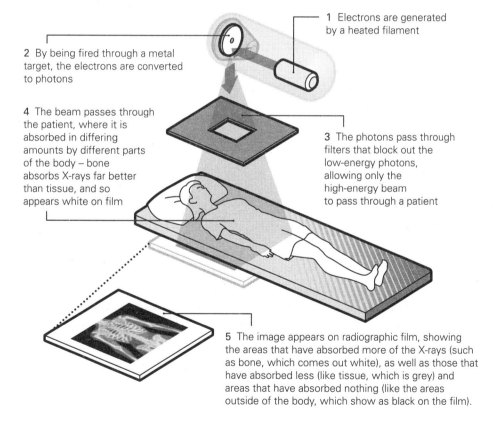

1 Electrons are generated by a heated filament

2 By being fired through a metal target, the electrons are converted to photons

4 The beam passes through the patient, where it is absorbed in differing amounts by different parts of the body – bone absorbs X-rays far better than tissue, and so appears white on film

3 The photons pass through filters that block out the low-energy photons, allowing only the high-energy beam to pass through a patient

5 The image appears on radiographic film, showing the areas that have absorbed more of the X-rays (such as bone, which comes out white), as well as those that have absorbed less (like tissue, which is grey) and areas that have absorbed nothing (like the areas outside of the body, which show as black on the film).

FUNCTIONAL MAGNETIC RESONANCE IMAGING

The key to revealing what's going on in our brains

Nominated by **Liz Bonnin**, presenter on BBC science show *Bang Goes the Theory*

Your body weight is made up of around 42 per cent muscle, 13 per cent bone and 16 per cent skin; most of the rest is water. X-rays are great at revealing bone structure, but it's the water in our bodies that is vital for magnetic resonance imaging (MRI) machines to draw up a picture of the soft parts, such as muscles, nerves and organs, and identify any abnormalities that shouldn't be there.

Each water molecule possesses two hydrogen nuclei, or protons. When an MRI machine is switched on, a giant superconducting coil generates an intense magnetic field, which causes the spin of some of these protons to align with the direction of the field. Pulses of radio waves are then used to flip the spin of the aligned protons in each water molecule in the body. When the machine is switched off, the protons revert to their original spin alignment. As they switch back, they release energy in the form of photons. The scanner detects all the photons and produces a 2D image. Layers of these 2D images are then stacked together to form a 3D image. Different tissues produce different photon patterns, because each tissue type contains a different amount of water.

The magnetic field strength is bumped up or down, depending on what part of the body is being analysed. In addition, 'contrast agents', such as the element gadolinium, can be injected to make blood vessels, inflammation and tumours stand out better.

The invention of the MRI scanner was revolutionary as it allowed doctors to investigate soft tissues safely, without exposing the body to the dangerous ionising radiation used in CT scans and traditional X-rays. The origins of MRI can be traced back to the 1930s.

Nobel Injustice

Physicist Isidor Rabi was born in what is now Poland, but his family moved to the US just a year after he was born. After graduating and gaining his PhD, Rabi relocated to Europe, where he worked with such eminent scientists as Niels Bohr. But it was on returning to the US that he made the biggest breakthrough of his career.

In 1930, Rabi started investigating the force binding protons together in atomic nuclei. Over the following years, he developed a technique called 'nuclear magnetic resonance' that allowed him to probe the nucleus of an element. This enabled him delve further into the atomic world and won him a Nobel Prize in 1944. While useful to physicists, the technique seemed to have little application in the everyday world – that was until an American physicist spotted its massive medical potential.

In 1971, Raymond Damadian wrote to the journal *Science* about using nuclear magnetic resonance to distinguish between cancerous tumours and normal tissue. After successfully patenting his technique in 1974, he carried out the first full body scan of a cancer patient in 1977. But he wasn't the only one to be using the technique.

A year before, British physicist Peter Mansfield and his team captured an image of the cross-section of a finger using a slightly different magnetic resonance technique. Mansfield also developed an algorithm that speeded up the image analysis process from hours to a matter of seconds. Back across the Atlantic, American chemist Paul Lauterbur was also working on plans for an MRI scanner. He claims he scribbled his first ideas for the scanner on a napkin while snacking on a sandwich, and then carried out his investigations at night when the local chemistry lab wasn't using their nuclear magnetic resonance machine. Lauterbur imaged all sorts of things, including clams his daughter had collected on the beach. But it was his imaging of beakers of water that was to prove crucial in the evolution of the scanner, as it is the water in the human body that is now key to MRI analysis.

Lauterbur's first paper to the journal *Nature* was rejected, but his persistence eventually paid off. In 2003, he won a Nobel Prize, alongside Pater Mansfield. Damadian was completely overlooked by the Nobel Committee for the part he played in the invention of the MRI scanner. When he heard about Lauterbur and Mansfield receiving their Nobel Prize, he was furious and took out a full-page advert in a number of the big US newspapers, with the headline "The Shameful Wrong That Must Be Righted".

Mapping the Brain

In the early 1990s, before all this Nobel Prize controversy erupted, a more specialised method of magnetic resonance imaging – functional

MRI, or fMRI – was developed which captured changes in real time, showing variations in blood flow in the brain or spinal cord.

"We're learning more and more from fMRI about the most fascinating and least understood organ in the human body," says Bonnin. "We can now image brain activity resulting from specific actions and emotions, investigate the brain's huge potential for plasticity, and even discover how its subconscious pathways operate."

Scientists can now create 'maps' of the brain and analyse not only our deepest thoughts but also our darkest desires, by invoking certain emotions in volunteers. For example, in a study by researchers at Northwestern University in Illinois, fMRI scans of volunteers watching erotic DVDs showed that the emotion of lust lies deep inside the brain in the primitive limbic system. Regions like the nucleus accumbens – involved in craving and pleasure – are also active, as is the prefrontal cortex, which is involved in decision-making and moderating correct social behaviour. While researching another one of the original seven deadly sins – envy – researchers at the National Institute of Radiological Sciences in Japan found the ventral striatum at the base of the brain lit up when volunteers were read passages about more successful people.

Doctors are also now able to peer into the brain using fMRI and discover more about medical conditions. "Not only is this technology giving us a better understanding of the inner workings of our brain and how intelligence might ultimately be measured," says Bonnin, "but it also improves the diagnoses of disorders of consciousness and neurodegenerative diseases such as Alzheimer's and Parkinson's."

BIOMEDICAL ENGINEERING

Technological breakthroughs for
the medical and biological worlds

Nominated by **Professor Ross Ethier**, professor of
bioengineering at Imperial College London and holder
of the Royal Society's Wolfson Research Merit Award

Biomedical engineering, as its name implies, means using engineering
in the fields of medicine and biology. Throughout history, the more
scientists discovered about the human body, the more they came to rely
on technology to understand its complexity.

"The idea of applying engineering to medicine and biology isn't
new, dating back hundreds of years," says Ethier. "But it's only become
recognised as a discipline in its own right within the last half century or
so. To understand the impact of biomedical engineering, you just have
to look at the technology in a modern operating theatre; it would be
unrecognisable to someone from a century ago."

Indeed, biomedical engineers design everything from surgical
devices, such as heart-lung machines, to imaging systems including CT,

NANOMEDICINE

Nanomedicine is a rapidly developing field of biomedical engineering. In biology, minute fluorescing crystals, known as 'quantum dots', are being attached to cells to track their movements in the human body. Likewise, so-called 'nanomagnets' have been developed that can be inserted into mouse blood cells, which allow detailed images of blood vessels in MRI scans.

ultrasound and MRI (see page 284), to implants such as artificial hips and intraocular lenses.

They're also at the forefront of some of the most cutting-edge fields of science. "Biomedical engineers are leading developments in areas such as regenerative medicine and nanomedicine," says Ethier.

STEM CELLS

The 'miracle' cells to cure diseases

Nominated by **Professor Mark Walport**,
director of the Wellcome Trust

Sea urchins make good lab rats because they produce thousands of eggs and the embryos develop rapidly. At the Marine Biological Station in Naples in the 1890s, German biologist Hans Driesch spent his time slicing up sea urchin eggs in an attempt to understand how they develop.

Scientists had already discovered that an embryo develops from a ball of cells (see page 211). However, they assumed that each cell in the ball was needed in order to create a whole organism, in other words that each cell developed into the corresponding part of the animal it had been pre-programmed to become. While analysing sea urchins, Driesch made a groundbreaking discovery that tore this theory apart.

Driesch found that if he divided embryo cells after their first cell division, each cell still developed into a complete sea urchin, meaning that each cell already contained all the information it needed to form the whole creature. These embryonic cells are now known as 'stem cells'. Mammals have two types of stem cell. Embryonic stem cells have the ability to become any cell type, while adult stem cells exist in relatively small numbers in many organs and tissues, where they await activation, becoming replacements for specialist cells.

The key role played by stem cells in repairing damage to our bodies, replacing worn out tissue for example, gives them huge potential in so-called 'stem cell therapy'. Blood-generating stem cells from bone marrow have been used for decades, and there has been some success in using them to generate new pancreatic cells for patients with diabetes. The wider use of stem cells is, however, still in its infancy. Scientists hope that one day they'll be able to replace any diseased or damaged tissues and cure all sorts of conditions from blindness to spinal cord injuries and diseases like Alzheimer's.

But, as often with medical science, it's becoming clear that the path to these breakthroughs is likely to be long and hard. Researchers believed that stem cells would be ideal for transplants as they could be taken from the patients themselves and thus avoid the problem of rejection. But, in 2011, researchers were shocked to discover that some kinds of stem cell trigger rejection.

On top of this, the anti-abortion lobby has argued against the ethics of stem cell therapy because human embryos are experimented on and then discarded. In fact, research takes place before the first cell division has taken place, and the massive potential benefits of the technique have convinced governments to back research. While Europe has supported stem cell research for many years, it was only in 2009 that Barack Obama lifted a ban on using US government money for it.

CLONING

From Dolly the sheep
to cures for medical conditions

Nominated by **Professor Andrea Brand**, professor of molecular biology at Cambridge University and the Wellcome Trust/ Cancer Research UK Gurdon Institute

Cloning animals isn't easy. We've been cloning plants for centuries, by simply cutting a leaf off one individual and planting it to grow into a whole new one. Most animals, however, reproduce sexually, making it incredibly hard to clone them. The process, which goes by the technical name of 'somatic cell nuclear transfer', involves extracting the genetic information from one cell in an animal, inserting it into an unfertilised egg that has been emptied of its own genes, and then making that embryo develop.

While this is no easy task, scientists have been cloning animals for over half a century. In 1952, Americans Robert Briggs and Thomas King succeeded in cloning embryos of a type of leopard frog. "Briggs and King performed the first nuclear transfer experiments and were able to generate tadpoles using nuclei from blastula stage embryos," says Brand. "But they were unable to generate tadpoles when they used nuclei from post-gastrula stage embryos."

Blastula cells exist in an embryo in the earliest stage of development, when the fertilised egg has grown into a hollow ball comprised of hundreds of cells. The gastrula forms after this when the embryo consists of two layers of cells. As Briggs and King weren't able to create tadpoles after the gastrula stage, they concluded that irreversible changes occur in the nuclei of cells as they differentiate.

But British scientist John Gurdon thought otherwise. He managed to generate tadpoles using nuclei from adult cells of *Xenopus laevis* – the African clawed frog. While none of these cloned tadpoles grew into adult frogs, his work showed that nuclei from fully specialised cells still contained all the genetic information required to make a whole new animal. "Gurdon showed Briggs and King were wrong," says Brand. "He became the first to successfully clone adult animal nuclei from adult cells."

Dolly the Sheep

The poster child of cloning was Dolly the sheep. Born on 5 July 1996, she became the first cloned mammal. Her creators, Ian Wilmutt and Keith Campbell, working at the Roslin Institute in Scotland, had taken a nucleus from the udder of a well-endowed Finn Dorset sheep (hence the name Dolly, after the large-chested country singer Dolly Parton) and inserted it into another stripped-out egg from a Scottish Blackface sheep. A jolt of electricity kick-started the embryo's development, and it was transferred back to the womb of a Blackface ewe.

But Dolly's successful birth was after a long line of failed attempts: 276 in total. Although she managed to give birth to six healthy lambs, when she was put down on Valentine's Day in 2003 at the relatively young age of six, she'd suffered more than her fair share of medical issues, including arthritis, obesity and a lung disease.

A whole host of mammals have since been cloned from cats and dogs to horses and mules. In most cases, they suffer from various medical conditions, such as larger than normal organs, obesity and problematic immune systems. It's thought this could be due to errors cropping up in the gene reprogramming process during the actual cloning procedure. But it could also be because cells think they're older than they actually are, since so-called 'telomeres' at the end of chromosomes which act as sort of 'cellular clocks', are often shorter in cloned animals. No-one knows for sure.

In view of all these problems associated with cloning, what's the point of it? Aside from the potential to clone endangered species, beloved dead pets, or artificial meat as food, cloned animals could have huge medical benefits.

Polly the sheep is less well known than her predecessor, but far more important for the medical community. The embryo from which Polly developed had been genetically modified to include the gene for blood-clotting proteins in humans. People suffering from haemophilia type B lack this protein and hence suffer from excessive bleeding. It's hoped that people with this condition will be able to be treated by drinking milk from clones like Polly.

Human clones

Cloning humans is currently banned in most countries. That didn't stop Clonaid – a group associated with the bizarre Raelian cult in Canada who believe aliens started life on Earth by cloning humans – from claiming in 2002 that one of their group had given birth to cloned 'Baby Eve'. It was dismissed as nothing more than a publicity stunt. And it's not just cults that have made such claims. In 2001, two fertility doctors, the Italian Severino Antinori and the Greek Panayiotis Zavos, claimed they had treated 10 infertile mothers using somatic cell nuclear transfer, but the clones were never seen and it was considered another publicity stunt.

Therapeutic cloning was legalised in 2001 by the British Human Fertilisation and Embryology Authority. This is where human embryos are created by somatic cell nuclear transfer, but then terminated before the embryo develops, although cells for research will have been removed before that. The benefits of this technique are numerous and far-reaching, including finding treatments for conditions such Parkinson's and Alzheimer's disease.

The World
of Psychology

FREUDISM

Disproving the greatest psychological dogma

Nominated by **Professor Richard Wiseman**, professor of psychology at the University of Hertfordshire, whose books include **Quirkology** and **Paranormality**

The name Sigmund Freud conjures up an image of an unorthodox, head-strong neurologist, who was obsessed with the unconscious and sexual desire. Freud's ego had no parallel: he, reputedly, would tell critics that there was no need for him to provide them with evidence, as what he said was true.

"The biggest breakthrough in the history of psychology was the evidence showing that Freudian psychology was wrong."

Richard Wiseman

Freud is also notorious for his colourful personal life. He took drugs and there were rumours that he had an affair with his wife's younger sister, Minna Bernays. Despite oral cancer in later life affecting his speech and hearing, eventually causing his lower jaw to rot away, he refused to quit his beloved cigars. He had over 30 operations and eventually his whole bottom jaw was replaced with a prosthetic one, which he called 'the monster'.

Possibly because he was so unorthodox, Freud is one of the most well-known neurologists of all time. He nevertheless deserves a place in the annals of history, as he was responsible for creating the field of psychoanalysis. "Freud was enormously influential in his day," says Wiseman.

Delving into the Subconscious

Sigismund Schlomo Freud was born on 6 May 1856 in Pribor, in what is now Czech Republic. Although his parents were poor, they made sure that their first-born received the best possible education. A bright student, Freud considered a career in law, but instead chose to study medicine at the University of Vienna.

While Freud agreed with Descartes that the essence of each person was down to their brain (see page 274), he thought the Frenchman had got one crucial thing wrong. It wasn't the conscious brain that determined who we are, but the subconcious – otherwise known as the unconscious.

After a brief foray into hypnosis as a form of therapy, Freud moved on to encourage patients to talk about repressed thoughts and emotions, attempting to make the unconscious become conscious.

"Freud tried to unlock the unconscious brain in a safe environment through therapy sessions, such as talking about dreams, and also through projective tests using ink blots," says Wiseman. In the ink-blot test, otherwise known as the 'Rorscach test' after its Swiss creator, patients' interpretations of ink blots are analysed using algorithms.

"There's been lots of research looking at these projective tests, and none of them holds any water," says Wiseman. "They simply don't tell you anything about the person in any systematic way. For example, the Freudian idea claims that you repress aggressive thoughts into the unconscious and the best thing to do is get them out in a safe way by punching something like a pillow. But again there's no evidence for that. In fact, if people behave in an aggressive way they feel more aggressive, not less."

Sexual Abuse

Another of Freud's theories was that personality and thoughts are a product of the child–parent relationship. Patients were encouraged to delve deep into their unconscious to extract repressed memories of sexual abuse. To account for mental disorders such as neurosis, Freud was reported to have put considerable pressure on patients to remember horrific instances from their childhoods, many of which might not have actually occurred.

"Freud claimed that everything you think and do is based on your relationship with your parents," says Wiseman, "but again the research just doesn't support that. As long as you didn't have a horribly abusive childhood, it really doesn't matter what sort of relationship you had in terms of what sort of person you are."

"I suspect a lot of Freud's incorrect theories may in part be due to his drug-taking. For many years his theories had a colossal influence in science and even more so in the arts. So the work that disproved Freud's flawed theories was the major breakthrough in psychology in the latter half of the last century."

Freud was a radical thinker, some of whose theories ended up being slightly wide of the mark. He was criticised by the likes of Carl Jung and Karl Popper because he often provided little or no evidence for his assertions. In fact, when Popper was developing his criterion of falsification to characterise scientific theories (see page 21), he singled out Freud's work as an example of something that is quintessentially unscientific. Popper's views rest on the assumption that there will never be ways to put the theory to the test. Of course new techniques, such as functional MRI scans (see page 284), can now help test Freud's theories. And, although it's early days, recent neuropsychological research suggests that Freud's dream theory was on the right track, as a close link has been found between the parts of the brain involved in dreaming and those responsible for motivations and emotions.

Many neurological theories remain unproven. Man has climbed the world's highest mountain, sent submersibles to the deepest part of the oceans, landed rovers on Mars and stepped foot on the surface of our lunar neighbour. But, despite garnering more and more knowledge about how our brain works, it continues to baffle us and a lot of questions remain.

ATTACHMENT THEORY

How understanding children's needs led to the redesign of hospitals and schools

Nominated by **Professor Simon Baron-Cohen**, professor of developmental psychopathology at Cambridge University and author of *Zero Degrees of Empathy*

"Children should be seen and not heard", the Victorians used to say. But in many upper-middle class families before the 1950s children often weren't even seen. Parents who spent too much time with their children were deemed to be spoiling them. Parenting duties were often delegated to a nanny, and many mothers saw their children only for an hour or so after tea-time.

"Attachment theory is extremely important, because it transformed our understanding of child development, education and the health service."
Simon Baron-Cohen

John Bowlby was like any other upper-middle class boy, being raised by a nanny and rarely seeing his parents. His mother was a social butterfly; his father was the surgeon to the King's household. Tragically, the much-loved family nanny died when John was four years old. Then, a few years later, he was packed off to boarding school at the age of seven. These childhood experiences meant that throughout his life Bowlby had huge empathy for the plight of neglected children, and steered his career towards researching the impact of poor child–parent relationships.

After studying psychology and pre-clinical studies at Trinity College, Cambridge, he worked with delinquent children before moving on to study medicine at University College Hospital and train in adult psychiatry at the Maudsley Hospital. In World War II Bowlby became Lieutenant Colonel in the Royal Army Medical Corps and, after the war, he worked as a Mental Health Consultant to the World Health Organization, and later he returned to psychiatry, becoming Deputy Director at the Tavistock Clinic in London.

It was during his time at the clinic that he developed his attachment theory on child–parent relationships. "While working at the Tavistock in the fifties and sixties, Bowlby came up with the idea that an infant forms an attachment to its care-giver," says Baron-Cohen. "The quality of that attachment is what determines the child's later emotional development."

"It's a simple idea, and it sounds very obvious because, today, we take it for granted that when parents have their first child they should be paying attention to the quality of the environment and relationship they're providing for them. But before Bowlby's attachment theory, very few people had ever articulated that."

Sigmund Freud (see page 299) had argued that events that happened during the first five years of life influenced personality development. Bowlby built on the foundations of Freud's idea, but explored in depth how the initial bond with the care-giver determines a child's later outcome.

Since Bowlby, numerous studies have been carried out on child–parent bonds. Austrian ethologist Konrad Lorenz researched behaviour in geese. The discovery he's probably most famous for is 'imprinting'. He found that the first moving object a gosling sees it deems its mother. Lorenz often found that goslings would 'imprint' on him, following him around as if he was their mother.

The infant–parent bond has also been studied through observations of rhesus monkey populations. Rhesus monkeys normally live in large social groups, called 'troops'. An infant normally spends almost all of the first few weeks in physical contact with its mother, then gradually starts to explore the local habitat. Every year most mothers leave their infants for a few hours or days to mate with males in the troop. Most infants react angrily at first, but then cope with the separation and appear normal when they're reunited. But highly strung infants, who show disruptive behaviour, cope poorly with the separation and cling to their mothers when reunited. So too in humans, some infants are more emotionally unstable than others.

Redesigning Hospitals and Schools

Before the 1950s, when children were taken into hospital they were separated from their parents and handed over to doctors and nurses. After Bowlby published his theory, children's hospitals were completely redesigned, so that parents could stay with their child if needed.

"This minimised the trauma of separation and breaking that attachment relationship," says Baron-Cohen. "Separation can have a damaging effect if a child is trying to cope with illness or a medical procedure and, on top of that, is also suffering from the absence of parental love."

Primary school classrooms were also completely reshaped in response to Bowlby's attachment theory. Parents had previously been expected to hand over their children at the school gates, and the children would then sit in rows of desks and obey an authoritarian teacher. After the theory was published, parents were welcomed into primary schools, and a more caring experience was created for children as they settled into their first school. A child's social development is now seen as just as important as its academic development.

There has been extensive research on attachment in humans. One conclusion is that children who don't get the chance to form a secure and loving child–parent bond are vulnerable to later mental-health issues and delinquency. As Baron-Cohen points out: "Children who have experienced deprivation, abuse or neglect are at a much greater risk of personality disorders in their teens and even in adult life."

The World
of Numbers

ZERO

The rise of nothingness

Nominated by **Timandra Harkness**, science writer and comedian

Not even the greatest mathematical minds of classical antiquity could conceive of zero as being a number. The credit for this must go to the Indian mathematicians, who, by the ninth century AD, had included the concept zero and its symbol 0 in their system of numeration.

This was important for two reasons. First, it enabled the quantity zero to be manipulated in calculations like any other number. (The only thing you can't do with zero is divide by it!) Second, a symbolic 'zero' is essential to any sensible method of representing numbers such as in our decimal system, where the value of a numeral is indicated by its position in the number.

"Zero is now so central to the way we think, and to our key technologies, that it's hard to imagine a time before we had it. For example, without zero, computers would only have ones."
Timandra Harkness

Compared with the Greek and Roman numerals, calculations were much easier to perform using the ten digits of the decimal system. The

Arabic world realised this and adopted the Indian numerals, including zero, for their computations. However, for many centuries, Europeans were having none of it. "When zero first reached Europe it was seen as foreign, suspect and even blasphemous," says Harkness.

Eventually the twelfth-century Italian mathematician Fibonacci – influenced by the writings of the Arab mathematician al-Khwarizmi – successfully introduced the decimal system and zero into Europe. Indeed, we now refer to our everyday numbers as 'Arabic numerals' and the word 'zero' itself comes ultimately from ṣifr, the Arabic word meaning 'empty' or 'void'.

EUCLID'S PROOF

The dawn of analytical thought

Nominated by **Professor Marcus du Sautoy**, Simonyi Professor
for the Public Understanding of Science and professor
of mathematics at Oxford University

Chinese students perform better in mathematics than their Western
counterparts. Part of this may be down to nurture: all undergraduates
have to pass tests in advanced trigonometry and algebra. But research
shows it could also be due to nature.

A recent study at the Dalian University of Technology using MRI
scans found that the Chinese volunteers used brain regions more
involved in the visual appearance of manipulating numbers, rather than
the Western volunteers who used the areas which assess the meaning
of words. So, it seems that Chinese speakers actually 'see' the numbers
better. Good news for the future of China, which is deemed to be the
next science superpower.

"Maths is crucial, because it's the language of science," says du Sautoy.
"You wouldn't start to build a bridge without doing the maths first. If
you want to understand what is happening in the Large Hadron Collider
at CERN, it comes down to analysing the maths that underpins all the

particles. If you want to monitor the spread of the swine flu virus, you need to use maths to predict what will happen and how you can control it. Mathematics is a fantastic language for looking into the future."

So, no surprises that du Sautoy considers the moment in science that truly changed the world to be when a mathematician – Greek, not Chinese – started thinking, out of the box.

Prime Example

The mathematician in question was Euclid of Alexandria. Details of his life are sketchy, but he's believed to have been born around 300 BC. His book *Elements* is considered to be one of the most influential in the history of mathematics. In it, Euclid covers geometry and perspective, but, importantly, he also discusses number theory and proves that there are infinitely many primes.

This might not sound like truly groundbreaking stuff that's going to affect our everyday lives. It doesn't exactly have any influence over the outcome of the Premier League or tell us whether we should take the brolly out today.

However, this was not just a pivotal moment in mathematics but also in science. "You might well say, who cares whether there are infinitely many primes or not?" says du Sautoy. "But this was the turning point when we started to realise that we can use analytical thought to access eternal truths."

The Babylonians and Egyptians were using mathematics way before Euclid's time, and starting to demand more abstract analytical thinking. "But this was the crucial step," says du Sautoy. "It produced certainty in a world filled with huge amounts of uncertainty; but also from a finite amount of logical argument the human mind had accessed the infinite."

"You could say all sorts of inventions or theories were the biggest moment in science. For example, Newton's calculus could be considered a turning point. But I think it is when mathematics begins. Euclid's book *Elements* really kick-started mathematics and modern science as well. This is when we changed the way we looked at the world. It underpins every single technological advance throughout the last few millennia."

Trailblazer(s)

Intriguingly, some scientists and historians believe that Euclid may not have existed at all. His life's work may have in fact been an accumulation of work by a group of people. But does it matter? Euclid's breakthrough in science from passive to analytical thought was a massive step, whether it was an individual or a group.

"For me it's not important," says du Sautoy. "That's the beautiful thing about mathematics: it stands up on its own without really having to know who did it or why they did it. It kind of speaks for itself."

CALCULUS

Discovering the mathematics of change

Nominated by **Professor Ulrike Tillmann**, professor of mathematics at the University of Oxford and fellow of the Royal Society

Where there's controversy in science history, often Isaac Newton is at the heart of it. This is the case in the story of calculus – the study of change.

Various ideas relating to precursors of calculus can be traced all the way back to the ancient Egyptians, and on to the ancient Greeks, such as Archimedes. In the fifth century AD, Chinese mathematician Zu Chongzhi used an early form of calculus to find the volume of a sphere. This became known as 'Cavalieri's principle' after the Italian mathematician with the interesting first name of Bonaventura, who used exactly the same method 1,000 years later. It wasn't until the seventeenth century, when two French mathematicians spotted a useful link between algebra and geometry, that calculus really came into its own.

René Descartes and Pierre de Fermat had independently worked out that the coordinates system used on maps could be adapted for use on graphs. Fermat's big breakthrough was to realise that one single equation could give any point along a curved line on a graph. These equations could then be manipulated using algebra. This doesn't sound groundbreaking in itself, but in those days the applications were enormous, from working out the distance covered by a rotating wheel to calculating the orbits of the planets.

Suddenly, a whole new branch of mathematics had been created where movement and change could be analysed. As to who invented true calculus is up for debate – enter the ever-contentious Isaac Newton.

Newton accused the German mathematician Gottfried Wilhelm Leibniz of plagiarism, claiming he'd stolen his ideas from unpublished notes that he'd shared with a select few at the Royal Society. The Society set up a committee to resolve the dispute – Newton was president at that time and wrote the final report.

However, if both mathematicians' notes are analysed in detail, it's obvious that they both arrived at the same conclusions independently. While Newton was the first to apply calculus to physics in general, Leibniz is now credited with creating a clear set of rules and providing much of the notation we use today. Leibniz coined the term 'calculus'; Newton called it the 'science of fluxions'.

"Despite the historical controversy over whether it was Newton or Leibniz who invented calculus first, there is little disagreement that calculus changed the world of science and, with it, the world itself," says Tillmann. "It empowers us to calculate and optimise, and gives us the tools and language to understand change, from the movements of the planets to economic change and even climate change."

THE FOURIER TRANSFORM

How an esoteric mathematical discovery unlocked a host of secrets

Nominated by **Professor Ian Stewart**, professor emeritus of mathematics and digital media fellow at the University of Warwick; author of numerous books, including *Professor Stewart's Cabinet of Mathematical Curiosities*

The French mathematician and physicist Jean Baptiste Joseph Fourier isn't as well known as some scientists in history, but his work has found applications in everything from photography to earthquake analysis.

Being orphaned at eight years old might have been a set back to someone less bright than Fourier. But, after a bishop took pity on him, he was educated at a convent, and ended up getting a lecturing job in mathematics. As a French revolutionary, he was rewarded with the chair at the École Polytechnique and a place on Napoleon's Egyptian expedition in 1798. And, after a stint as governor of Lower Egypt and secretary of Institut d'Égypte, he returned to France where he started working on heat transfer.

While tackling some problems concerning the flow of heat through objects, Fourier hit on the idea of breaking down quantities that vary in complex ways into very simple components made up of waves. Fourier claimed that he could represent any such complex variation by a combination of mathematical functions called 'sines' and 'cosines'. While his claim took many years to prove beyond doubt, what is now called 'Fourier analysis' was quickly put to use, helping to solve tricky problems by breaking down quantities that vary in space or time, or both, into combinations of sines and cosines.

A key part of Fourier analysis is the breaking apart of a complex varying quantity, such as a radio signal, to find its constituent waves. This involves calculating so-called 'Fourier transforms', which will reveal the key characteristics of the waves making up the signal. This has made Fourier's techniques immensely valuable in fields as diverse as crystallography, optics, signal processing and geophysics. "It allows the structure of large molecules to be found, especially those occurring in biology, using X-ray diffraction," says Stewart. "It also enables compression of image data in digital photography, making it possible to store more images on a card of given capacity. It makes cleaning up old or damaged audio recordings possible, and on a grander scale enables earthquake data to be analysed. Those are just some of the uses."

PROBABILITY THEORY

Laying down the laws of chance

Nominated by **Robert Matthews**, visiting reader in science at Aston University and science consultant at *Focus* magazine

"It's ironic that we live in a world full of uncertainty, risk and random events, and yet most of us have no idea about the laws governing such things, namely the theory of probability. It's also amazing just how long it took mathematicians to pin down these laws – especially given that people have been gambling for millennia!"
Robert Matthews

No-one's really sure why it took so long to develop the laws of probability theory. It could be that chance events were seen as acts of God, and so no attempt was made to try to fathom what He was up to. But Matthews says part of the blame may lie with the ancient Greeks – and one man in particular.

The fourth-century BC Greek philosopher Aristotle developed classical logic, which focused on black and white issues producing true or false outcomes. In reality, most things we worry about in life involve shades of grey. "This is one of the great strengths of probability," says Matthews. "It has the power to reflect this fact about our world, as it gives its answers as a number between zero (impossibility) and one (certainty). It's a far more powerful means of understanding the world around us."

The Laws of Gambling

In 2010, the UK betting industry was worth £6 billion. That's a sizeable amount of money and shows just how lucrative it can be to have a really good handle on the rules governing chance and uncertainty.

Money appears to have been the primary motivation behind the question that sparked the systematic study of the laws of probability. The seventeenth-century French writer Antoine Gombaud liked to get into the characters that he wrote about in his books. So he adopted the name Chevalier de Méré, despite not being a knight. The nickname stuck and his friends started calling him that. He was also an amateur mathematician, who enjoyed playing around with numbers – and making bets. During one gambling session, he came across a problem which, though seemingly esoteric, marked a turning point in the development of probability theory.

Imagine two players decide to play a number of games, for example the best of ten tosses of a coin, with the aim of winning a certain pot of money. But for some reason they can't finish all the games. How should the stake be split if, say, someone's won four games and the other only three? To solve this, Chevalier de Méré got in touch with the brilliant French polymath Blaise Pascal. He took up the challenge and together with French lawyer Pierre de Fermat (of Last Theorem fame) they lay the foundations of probability theory.

So, nowadays, gambling is not just a game of chance: knowing a bit of probability theory can help you make money, or at least maximise your chances of doing so. For example, imagine you go into a casino and you have £100 to spend. You pick the simplest game: choosing between red or black in roulette. If you know anything about casinos, you'll know that the odds are deliberately chosen to be unfair, so that the house always makes a profit in the end. So what should you do: have ten goes at £10 a bet or just chance the whole £100 in a single go?

"Probability theory reveals the answer: go for broke, and bet the lot in one play," says Matthews. "You're substantially more likely to come out ahead that way. And even a slight understanding of probability will explain why: with cautious play, you're giving the casino more opportunities for its unfair odds to rob you blind."

Serving Justice

For non-gamblers, probability theory has numerous other applications, such as finding out what to believe about evidence.

In the 1920s, the English polymath Frank Ramsey showed that probabilities don't just help with chance events: they also capture the key idea of 'strength of belief'. So, for example, if you're pretty sure about something, you can say your level of belief is close to 1 (certainty), or if you're unsure your belief is closer to 0.5. Ramsey showed that a specific theorem in probability theory, known as 'Bayes's rule', can be used to update your belief about something in the light of evidence. Put simply, the rule says that you should take your original (prior) level of belief and multiply it by a factor that reflects the weight or strength of evidence that you've just come across, something called the 'likelihood ratio'. The result gives you an updated level of belief.

"This application of probability theory is incredibly useful," says Matthews. "It can prevent miscarriages of justice by ensuring juries are not being overly swayed by unconvincing evidence. And it casts light on a host of other issues. For example, it shows why earthquake predictions are – and always will be – rubbish. This is because no prediction system can ever supply the weight of evidence to compensate for the fact that big earthquakes are inherently rare. It even makes sense of such controversies as the flaws in animal experiments: the problem is that no-one has actually shown that, for example, animal tests of new drugs provide useful weight of evidence about what will happen with humans."

Weird Coincidences

Probability theory isn't all about useful applications, though. It also yields some truly bizarre insights and throws up a whole host of astonishing counterintuitive results. "There are so many of these," says Matthews. "For instance, you need just five people in a room to have a much better than 50:50 chance that at least two share the same star sign, or that you were born in the same month. Or that just 23 people are enough to give better than even odds that at least two share the same birthday."

From predicting coincidences to avoiding injustices to helping us make money, probability theory is astonishingly versatile. Asks Matthews: "How many other areas of maths or science do you know that can do all this?"

PUBLIC-KEY CRYPTOGRAPHY

Keeping our lives secure

Nominated by **Dr Simon Singh**, physicist and author whose books include *Fermat's Last Theorem* and *The Code Book*

For all its hi-tech image, data security dates back millennia. A small clay tablet covered with a code written in Mesopotamian cuneiform and found on the banks of the Tigris River has been dated to around 1500 BC. The use of ingenuity to conceal messages is similarly ancient: secret messages were tattooed on the shaved heads of slaves in Ancient Greece, so that the message would be concealed from enemy eyes as their hair grew back, until their head was re-shaved by the person receiving the message. That worked well most of the time, but the recipient had to know which slave's head to shave. Plus, if the enemy cottoned on to the scheme, by process of elimination and with a knife in hand, they could find out the message themselves.

Several Middle Eastern civilizations had already devised a more sophisticated system by which written messages could be encrypted by the sender and decrypted by the recipient. These worked by simply substituting letters of a plaintext according to a pre-agreed system, to create what is known as a 'ciphertext' – an encoded version of the original message. But if someone worked out what letters had been substituted, they could easily break the cipher.

In the ninth century, Arab polymath Al-Kindi dealt such methods a severe blow by putting forward various other ways of cracking ciphers – including the 'frequency analysis' method, which worked on the basis that some letters tend to be used far more often that others; for example 'e' makes up around 12 per cent of English text. Thus by analysing the more frequently used letters in the ciphertext, it's possible to deduce the plaintext letters. One way to counteract this is to use different cipher letters to represent the same plaintext letter. While the creation of such a 'polyalphabetic' cipher is usually credited to a fifteenth-century Italian scholar, there's some evidence the technique was known to Al-Kindi.

Over the centuries, ever-more complex cipher systems have been devised and ever-more sophisticated techniques have been employed to break them.

During World War II, the Germans adopted a commercial encryption method they believed to be unbreakable, based on the Enigma machine. What they didn't know was that, before the war had even started, mathematicians working at the Polish Cipher Bureau had already found ways of undermining the Enigma machine's security. The Bureau passed that information on to French and British code-breakers, who built on the original research to develop methods for reading many of the messages almost as soon as they were transmitted. When the German High Command introduced a still more sophisticated encryption machine, the code-breakers developed even more sophisticated methods – including the invention of the first programmable electronic computer – to break its messages.

So it seemed that no cipher system, however complex, was unbreakable, as the recipient had to have some prior knowledge of the system being used. "If I'm going to scramble a message, I have to tell you how to unscramble it," says Singh. "I can't tell you on the phone because someone might be listening. I can't send it to you in the post because somebody might intercept it. So we have to meet, in order for me to give you a code to decipher the message. But that completely defeats the object, because I might as well give you the message there and then."

But in the 1970s, two Americans devised a solution.

Under Lock and Key

In 1976, Whitfield Diffie and Martin Hellman published a ground-breaking scheme. With their key exchange system, two people who didn't know each other could exchange a secure message over an insecure communication channel.

A good analogy of this 'public-key cryptography' is a box with a secret object inside. If the box is padlocked and sent to another person, the recipient can't open it because the sender hasn't given them the key. But if the recipient puts their own padlock on the box and posts it back to the sender, the sender can unlock their padlock, send it back to the recipient who can take their padlock off, because they've got that key.

It's a fairly simple idea. Unlike other technological advancements, such as travelling to the Moon, which needed the development of the rocket, or supercomputers, which needed microchips, public-key crytography could have existed with early twentieth-century technology. Society just didn't need it to the same extent a hundred years ago. "We now live in the Information Age, where information is precious. Public-key cryptography is one way to safely store information in databases or communicate it securely," says Singh. "I don't think it is the most important development in all of history, but it's crucial in the twenty-first century because we live in this Information Age."

We now know that Diffie and Hellman weren't the first to use decryption algorithms. At the end of the twentieth century, rumours spread that the Government Communications Headquarters (GCHQ) in the UK had independently developed a similar system. In 1997, GCHQ came clean about the fact that researchers James Ellis, Clifford Cocks and Malcolm Williamson had developed a system called 'non-secret encryption' in 1973. "They kept this secret, because they wanted to protect this code-breaking technology," says Singh. "I don't know to what extent they were using it, probably just among their own secure network. But we just don't know."

Quantum Future

In 1977, Ron Rivest, Adi Shamir and Leonard Adleman from Massachussetts Institute of Technology made a big breakthrough in public-key cryptography, by developing an algorithm that made Diffie–Hellman work in a practical way. This algorithm was called 'RSA' after the surname initials of its developers. Without it, we wouldn't be able to carry out financial transactions such as shopping online or using pay TV, we couldn't detect forgery or tampering, or have secure mobile phone calls.

The war against code-breakers never ceases, and now the laws of the sub-atomic world are being brought in to help keep messages secure. The technique relies on the fact that any attempt to eavesdrop on such 'quantum communication' methods – such as photons of light prepared in special ways – is instantly revealed. "The recipient measures the photons using polarised filters, which are orientated in a particular way," says Singh. "The recipient doesn't know if they're measuring them correctly, but afterwards they tell the sender how they measured them, and the sender can then tell them if that was correct."

"Spinoff companies are developing new quantum computing technologies for very secure financial transactions," says Singh. "If anybody intercepts the photons, the recipient can tell that the photons have been disturbed and the message has been tapped. It's no longer the science fiction future, it's very real."

A good analogy of this 'public-key cryptography' is a box with a secret object inside. If (1) the box is padlocked and sent to another person, the recipient can't open it because the sender hasn't given them the key. But if the recipient (2) puts their own padlock on the box and posts it back to the sender, the sender (3) can remove their padlock, send it back to the recipient who can take their padlock off, because they've got that key.

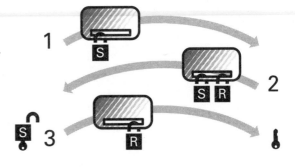

In reality each person owns a pair of keys which work together. Their public key is shared openly and used to encrypt incoming emails. The incoming mail can only be decrypted by their private key.

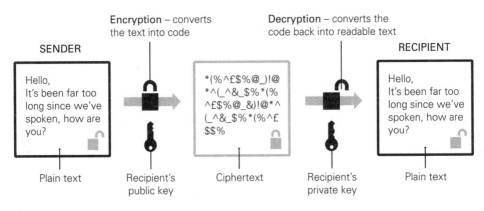

Encryption – converts the text into code

Decryption – converts the code back into readable text

SENDER

RECIPIENT

Hello,
It's been far too long since we've spoken, how are you?

(%^£$%@_)!@ *^(_^&_$%(% ^£$%@_&)!@*^ (_^&_$%*(%^£ $$%

Hello,
It's been far too long since we've spoken, how are you?

Plain text

Recipient's public key

Ciphertext

Recipient's private key

Plain text

The World of Engineering

THE WHEEL

The invention that changed transport – and revolutionised machinery

Charlie Turner, editor-in-chief
of *Top Gear* magazine

Almost every machine with moving parts has wheels inside, from disk drives to cars.

While logs have been used to move objects for thousands of years, no-one knows exactly when the first wheel was invented or what it was used for. We do know, however, that they existed before 3500 BC in ancient Mesopotamia, as a potter's wheel has been discovered which dates from around that time.

The oldest known transport wheel and its axle were discovered in 2002 in a boggy marsh near Ljubljana in Slovenia. Wood doesn't preserve well, but the muddy water had protected it from the rotting agents of bacteria and fungi. Radiocarbon dating puts it between 5,100 and 5,350 years old.

Evidence from diagrams on clay tablets suggests that wheels for transport didn't catch on for a while, though. This could be because oxen did a perfectly good job of lugging ploughs around, while all sorts

HISTORY OF THE TRANSPORT WHEEL

3300–3000 BC	First transport wheels made from solid wood and connected by an axle.
3000 BC	Hard rims fortified with metal strips and nails were introduced that made wheels more durable.
2600 BC	Thinner, lighter planks gradually replace solid wood wheels, which made chariots slow and cumbersome.
1600 BC	Egyptians invent the spoked wheel.
800–600 BC	The Celts invent pivoting front axles which make vehicles more manoeuvrable.
1400–1500	The first 'tyres' – iron bands – reinforce wheel rims.
1820s	Invention of metal hubs, enabling heavy steam vehicles to move without snapping spokes.
1846	Robert William Thomsen patents the pneumatic tyre: an inflated belt of rubber.
1967	Invention of alloy wheels which are lighter than steel and conduct heat away from the brakes more effectively.

of animals from camels to yaks carried humans. But it could also be down to a Catch-22 scenario.

Wheels need smooth surfaces to roll on in order for the vehicles they carry to move both quickly and efficiently. But the problem was that roads able to handle fast-moving chariots weren't going to be constructed until there was plenty of demand for them. In the end, roads did evolve to take speedier chariots, but this Catch-22 re-emerged a few centuries later when wheel and vehicle design didn't take a leap forward until modern road design arrived..

Frenchman Pierre Marie Jérôme Trésaguet brought science to road construction in the mid-1700s. His designs used curved edges and drains on either side. But, crucially, he came up with the idea of having a base layer of large stones covered with a thin layer of smaller stones. This clever idea meant that when rolled on by traffic, the stones jammed together forming a stronger surface.

Scotsman John Loudon McAdam improved on this design when, in the 1820s, he mixed compacted broken stone aggregate with a cementing agent aggregate to form a strong, durable surface, which he named 'macadam' after himself. Then, in 1901, Edgar Purnell Hooley patented his invention of 'tarmac': a tough tar and aggregate mix. As road materials improved, wheel design took off and vehicles got faster and faster.

THE SCREW THREAD

The simple device that cranked technology to new levels

Nominated by **Jem Stansfield**, engineer and presenter on BBC science show *Bang Goes the Theory*

More screw threads are produced each year than any other machine component. That's because they are so integral to almost every bit of technology in the world around us. Even if an object doesn't contain a screw thread, the machine that made it will have had one.

The screw thread is so-called because the ridge spiralling down the body of the screw is known as the 'thread'. The distance between the sections of the ridge that run parallel to one another – the 'pitch' – determines the screw's holding power. So the smaller the pitch, the more friction is created to hold an object in place and prevent it slipping.

"If the whole of something is ever going to be greater than the sum of its parts you've got to join those parts together," says Stansfield. "The screw thread is like driving a wedge under something, but then by wrapping the wedge round into a circle you can just keep on driving it with a simple turning action."

A clever invention, indeed, though tracing its origin isn't easy. It may have been invented around 400 BC by Archytas of Tarentum, in Magna Graecia, an area of southern Italy which was then under Greek control. Certainly, the idea behind the screw was used in oil and juice presses in ancient Pompeii and also in Archimedes' screw, a type of water pump said to have been invented by the eminent Greek mathematician Archimedes in the third century BC. In fact, this device – essentially a helical screw inside a hollow tube – had been used earlier by the ancient Egyptians for drawing water up from low-lying areas to irrigate their crops. Some scholars believe that a type of Archimedes' screw had been used even earlier by the sixth-century BC Babylonian King Nebuchadnezzar II to water the Hanging Gardens of Babylon, one of the original Seven Wonders of the World.

"The humble screw thread may simply be an inclined plane wrapped around a cylinder, but without it, metaphorically and literally, the wheels would come off the modern world. Nature barely uses the concept, but for mankind it holds false teeth in place, focuses microscopes, and fastens hatches on to spacecraft. Screw threads also provide the precision movement that allows the automated manufacture of just about every piece of technology we take for granted. Without the screw thread our world would be a very different place."

Jem Stansfield

Through the centuries, the metal screw was used in all sorts of technologies. However, there was a problem with screw components that needed to fit together, such as nuts and bolts: only a pair with an identical thread could be guaranteed to fit. At the turn of the nineteenth century, a young blacksmith had an idea. Previously, in order to make a screw, a cutting tool was held against the iron as a device known as a 'treadle' rotated, but the process wasn't precise. In 1800, Henry Maudslay invented the screw-cutting lathe, which had a toolholder into which the cutting tool could be clamped, allowing engineers to cut identical screws.

One of Maudslay's trainees was a talented mechanic called Joseph Whitworth. Under Maudslay's tutelage, Whitworth developed numerous precision machine tools, which meant that individual engineers didn't need to make their own tools, as they could now be mass-produced. Whitworth also developed a number of standard sizes for gauges and components, which were adopted nationwide by railway companies and the like. His legacy is remembered in standard screw sizes such as British Standard Whitworth. "Whitworth's influence on the world of engineering is pretty huge," says Stansfield.

As technology has developed, the screw remains the best way to temporarily attach objects together. "Welding joins objects permanently, whereas screws are adjustable," says Stansfield. "Adhesive manufacturers may claim to have made nails redundant, but 'No-more-nuts-and-bolts' is a product I can't see happening. Well over a thousand years after its invention the screw thread is still the best overall fixing method we've got."

THE FLUSH TOILET

How the invention of the loo and
mass sanitation cleaned up disease

Nominated by **Dr Henry Gee**, senior editor of
the journal *Nature*

The world's most expensive toilet reportedly cost millions of dollars. It's not made of bone china and encrusted with diamonds. The reason it's so pricey is that it's designed for a Space Shuttle.

Jettisoned rocket parts and other space junk that orbit the Earth are a massive problem. If a satellite, spacecraft or space station gets hit by even a tiny fleck of paint, at 10 km per second it creates a 1 mm-deep dent. Imagine what a doughnut-sized lump of excrement hurtling towards a space-walking astronaut would do.

So ejecting the contents of a toilet into space is not just poor bathroom etiquette, it is extremely hazardous. Instead, solid waste is stored and returned to Earth – a fuel-hungry

"Before the invention of the flush toilet and, by extension, the sewerage system, people were literally living in their own effluent and thus subject to all kinds of diseases, whether infectious or parasitic."
Henry Gee

load to carry – while urine on the International Space Station (ISS) is recycled into drinking water. This saves transporting water up to the ISS at a cost of $40,000 per gallon. That's quite a saving, but the cost of the spacecraft toilet must still seem an incredible extravagance to much of the world's population who don't own a toilet, or even a bucket, and have to 'go' out in the woods or over a drain in a crammed city street.

"Today, four out of ten people have no access to toilets of any kind and defecate on open ground," says Gee. "Every minute four children die from diarrhoea, 90 per cent of which is caused by exposure to untreated waste."

Faeces are hotbeds of disease. Just 1 g of faeces can contain 10 million viruses and 1 million bacteria, not to mention the 1,000 parasite cysts and 100 worm eggs. And the Earth's population is ever-increasing – it's now passed the 7 billion mark, and is set to reach 9 billion by 2045. By the time you finish reading this sentence, there will be 12 more people in the world than when you started. With facts like these in mind, it's maybe not surprising that, in 2007, *British Medical Journal* readers voted sanitation the biggest medical breakthrough in the last two centuries.

The Loo through the Ages

In fact, toilets have been around for centuries. Way back, around 5,000 years ago, in the Indus Valley in Pakistan, Bronze Age settlements had sophisticated sewage systems. No-one knows for sure, but it's likely that rudimentary loos existed above the systems, with a hole that people poured water through to flush away waste.

As with so many inventions, the Romans were ahead of their time too. Evidence of public toilets has been found in archaeological sites like Ostia Antica, an old harbour city near Rome. In ancient Rome, going to the bathroom was a social occasion – there were no walls or screens for privacy.

After the collapse of the Roman Empire, toilet technology came to a bit of a standstill. Archaeologists did discover a 2,000-year-old toilet that seemed to have a flush mechanism in the tomb of a Han Chinese king. But the first modern flush toilet wasn't invented until 1596.

Sir John Harington's Ajax model used weights and levers to pour water from a cistern, and then a valve opened to allow the water to flow out. This flush toilet was only available to the rich, and the development of sanitation for the British masses took a while longer to develop.

In 1849, 50,000 people nationwide died from cholera. As victim's bodies were carted out of the cities, people avoided contact with others for fear of infection. But what they didn't do was stop eating – and, crucially, drinking water.

In 1854, Dr John Snow decided to investigate the epidemic. He was sceptical of the popular theory of the time that noxious vapours filled with particles of decomposed matter (miasmata) caused illnesses such as chlamydia or cholera. Snow spent time talking to residents of London's Soho district and documenting their behaviour. Gradually, he saw a pattern forming.

Anyone who became ill with cholera was sourcing their water from a pump in Broad Street. He realised it must be contaminated. Although he couldn't prove the water contained excrement, laced with the agent of cholera, he succeeded in convincing the council to remove the handle of the water pump. The cholera epidemic in Soho died down.

The Great Stink

Snow's discovery was a catalyst for reform. The Thames had become an open sewer. After the Great Stink in the summer of 1858, when the smell of untreated sewage became unbearable, the government ordered a complete overhaul of the rudimentary sewage system.

Civil engineer Joseph Bazalgette was tasked with designing a comprehensive underground sewage system, diverting waste into the Thames, but downstream of London's core population.

By 1875, it was illegal to build a new home without a toilet. As with any invention, when government legislation changes, sales take off. A number of inventors were designing flush toilets by then, but in the 1880s Thomas Twyford designed the first toilet made out of one whole piece of ceramic. In just five years he sold an incredible 100,000 of them.

Millions of toilets are now sold every year, but there are still 2.5 billion loo-less people around the world. Every year the World Toilet Organization holds an international conference, which aims to break the taboo of talking about human waste and find solutions for those without a toilet or sanitation. This sounds like a pretty amusing event, but remember that, since mass sanitation was introduced, it has been claimed that the average person's lifespan has increased by 20 years.

THE NEWCOMEN ENGINE

The catalyst for the Industrial Revolution

Nominated by **Adam Hart-Davis**,
presenter and writer, whose books include **Science**

Coal is dirty stuff. But this 'black gold' has been used as a fuel for centuries, since around 300 BC. By the end of the Middle Ages, most of the open-cast mines were depleted and miners were having to dig deeper and deeper to keep up with the demand. The trouble was, the deeper the mines, the more prone they were to flooding. Enter Thomas Newcomen.

> "The Newcomen engine was the first effective steam engine, and the first source of portable power. It started the Industrial Revolution, and it changed the world forever."
>
> Adam Hart-Davis

The ironmonger grew up in Dartmouth, Devon, an area littered with tin mines. However, it was for a deep coal mine in Dudley that, in 1712, Newcomen built an 'atmospheric engine' to replace the manual and horse-driven pumps which had previously been used to extract water from the depths. The engine burnt coal to boil water and create steam, which then drove a piston.

HOW THE NEWCOMEN ENGINE WORKS

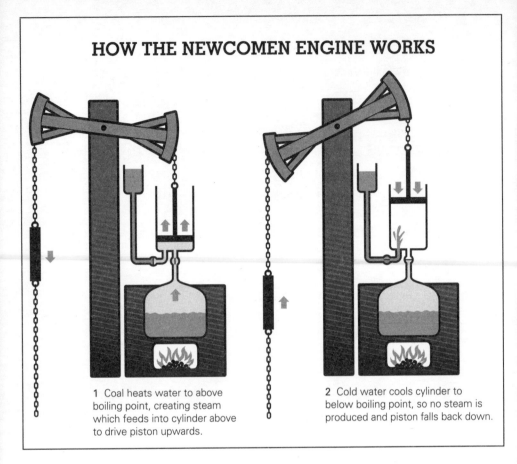

1 Coal heats water to above boiling point, creating steam which feeds into cylinder above to drive piston upwards.

2 Cold water cools cylinder to below boiling point, so no steam is produced and piston falls back down.

The engine was rugged, reliable and so popular that 1,000 were built throughout the eighteenth century and exported across Europe. But the engine was also inefficient, requiring lots of coal and wasting a shed-load of heat. In addition, each one cost £1,000 to build – a small fortune in those days.

When repairing a model Newcomen engine in 1764, Scottish engineer James Watt realised how inefficient it was. For every stroke of the piston, the whole working cylinder had to be heated to above and then cooled to below the boiling point of water. While walking on Glasgow Green, Watt had a eureka moment: "I had not walked further than the golf-house when the whole thing was arranged in my mind." If he could keep the cylinder constantly above boiling point and have a separate condenser, the engine would be far more efficient.

It took Watt a decade to build an engine that worked. But luckily he had the entrepreneur Matthew Boulton to back him, and for the next 25 years it was the engine of choice around the world, revving the Industrial Revolution up a gear.

STEAM LOCOMOTION

A new lease of life for the Industrial Revolution

Nominated by **Professor Seth Shostak**, senior astronomer at the SETI Institute and the author of **Confessions of an Alien Hunter**

"The development of practical steam locomotion caused a discontinuity in the history of humankind, marking the end of 10,000 generations in which lives were incurably dull, short and brutish. No longer and never again would we all have to stay poor and stay put."
Seth Shostak

James Watt's steam engine ruled factories around the world in the late 1700s. Its clever design of a separate condenser eliminated the constant heating and cooling of the cylinder, making it a lot more efficient than its predecessor, the Newcomen engine (see page 339).

It was only a matter of time before someone came up with the idea that the energy from a steam engine could also be harnessed to create forward motion. It's still up for debate as to who created the first steam vehicle. Nicolas-Joseph Cugnot built a three-wheeled steam wagon that transported cannon barrels and the like. But credit for the first roadworthy steam locomotive normally goes to Richard Trevithick.

The Cornishman built his 'Puffing Devil' in 1801, and on Christmas Eve of that year it crawled its way from one village to the next, carrying several men. Despite it breaking down three days later and going up in flames, Trevithick took out a patent for the machine, and a series of improved versions followed. Unfortunately, the lack of comfort and the fact that they were more expensive to run than horse-drawn carriages made them fairly impractical. A new application was needed.

In 1804, Trevithick's newest engine was challenged to haul a 10-tonne load of iron along the Penydarren Ironworks wagonway in south Wales. It exceeded expectations and on 21 February the machine made the 16 km journey. Trevithick had finally found a use for steam locomotives. The man was a genius, but he wasn't the best businessman, ending up bankrupt. While he was away, travelling around South America, more business-savvy inventors were hard at work improving on his creation. One such, was engineer George Stephenson – the 'father of the railways'.

As steam locomotives evolved, so too did the tracks on which they could run. Early wagonways made of wooden planks carried goods wagons between mines. As steam locomotives got heavier and heavier, the wooden planks weren't able to support so much weight. After numerous incidents with heavy locomotives damaging wooden tracks, an alternative was needed. Cast iron rails were still too brittle, but Stephenson started experimenting with tougher wrought iron rails and also with more locomotive wheels to distribute the weight more evenly.

In 1820, Stephenson was hired to build a 13 km stretch of railway from Hetton Colliery to Sunderland. Upon completion, it was the first railway not to use any animals for power. Stephenson was to gain even more recognition for his revolutionary superfast locomotive.

The aptly-named *Rocket* had 25 copper tubes running the length of the boiler to get rid of hot exhaust gases and produce more steam. In 1829, Stephenson entered it in the Rainhill Trials in Liverpool. Despite pulling a load three times its weight, it reached a speed of 19 km/hr. On another run, towing a carriage of passengers, it peaked at 39 km/hr, faster than a horse, an accolade which no vehicle had ever achieved before. The *Rocket* won the trials hands down and went on to see 67 years of service.

At such speeds, long-distance travel became a viable option for many, while mining had a new lease of life as coal could be transported more quickly around the country. Indeed, the steam engine dominated industry and transportation for 150 years.

DID YOU KNOW?

- The 'choo-choo' sound is created when the valve that opens the cylinder releases steam at high pressure, making a 'choo' noise.

- The fastest ever steam engine was the *Mallard*, which reached 204 km/hr.

THE CAR

How the automobile evolved

Nominated by **James Caan**, entrepreneur
and ***Dragons' Den*** panellist

A century ago, the automobile was a novelty. Today, around 700 million cars speed along our roads. Many of these are made by European manufacturers Mercedes Benz and Daimler and US manufacturer Ford, which were all founded in the late nineteenth or early twentieth century by their namesakes, Karl Benz, Gottlieb Daimler and Henry Ford. These men played a key role in the car's evolution, but the original idea behind the combustion engine that powers most cars dates way back to the thirteenth century.

> *"The automobile revolutionised transport and it was Henry Ford – the ultimate entrepreneur – whose incorporation of the assembly line into car manufacturing really transformed the industry."*
> James Caan

A tiny amount of high-energy fuel, trapped in a small, enclosed space and ignited, will release an enormous amount of energy as the gas expands. If this is done hundreds of times a minute, this energy can be harnessed to move a piston up and down. Attaching what is known as a 'crankshaft' converts this linear motion to rotation. In his *Book of Knowledge of Ingenious Mechanical Devices*, published in 1206, the Arab inventor Al-Jazari gave the first known description of a crankshaft.

Exactly 600 years later, the Swiss inventor François Isaac de Rivaz included a crankshaft in his working model of a combustion engine, which was powered by a hydrogen/oxygen mix. By placing this engine inside a wooden frame with wheels, he created what became known as an 'automobile'.

While de Rivaz's invention was promising, it flopped commercially. Nikolaus Otto's four-stroke engine was more successful as it burnt fuel efficiently, but it only performed back-and-forth motion. The man who turned that into circular motion was German engineer Karl Benz.

Unlike previous automobiles, which were simply motorised horse carriages or stage coaches, Benz's three-wheeled Motorwagen, built in 1885, was the first to generate its own power. He started selling these in 1888, but the car had no gears and couldn't climb hills without a bit of help, and, as there were no petrol stations back then, owners had to buy the fuel from pharmacies, which sold small amounts of it as a cleaning product.

After Bertha Benz suggested to her husband that he add another gear, she set out on the first long-distance automobile trip. On the 106 km journey from Mannheim to Pforzheim to visit her mother, Bertha refuelled at pharmacies, repaired mechanical problems and invented brake lining by asking a shoemaker to nail leather onto the brake blocks. Today, her route is marked as the Bertha Benz Memorial Route, and every two years an antique automobile rally retraces her tracks.

RUDOLF DIESEL

Designing a competitor to the petrol engine almost killed engineer Rudolf Diesel. Although his diesel engine didn't need a spark to ignite, in 1894 one of his prototypes exploded and he was hospitalised for many months. Having survived this near-death experience, he continued to suffer from health and eyesight problems, but by the end of the 1890s he'd developed his diesel engine, which made him a millionaire.

Diesel engines work by compressing air, then injecting fuel into the engine. They compress the air at a ratio of up to 25:1, compared with the compression ratio of around 10:1 in petrol engines.

Around the same time, Gottlieb Daimler invented the high-speed petrol engine and added an extra wheel to make the world's first four-wheeled automobile. But the man who made it affordable for the masses was Henry Ford.

American inventor Ford recognised that the car was so expensive because it took so long to build. With the advent of the industrial revolution, engineers realised the potential for speeding up the manufacturing process by getting different workers each to repeat their tasks over and over again. Ford spotted that this so-called 'assembly line' could be applied to car manufacturing. The framework of the car moved along a line of workers, who each added a new part. As the workers didn't have to move around the factory and became expert at adding a specific part, it only took around an hour and a half to build each car (all of which were painted black because it dried quickest) compared to other manufacturers taking around 12 hours.

Henry Ford was a clever boss: he paid his workers enough for each of them to be able to afford one of the cars that they were working on. The Ford Motor Company's first car went on sale in 1908. Almost two decades later, this Model T (aka the 'Tin Lizzie') was still in production, and, in its lifetime, over 15 million models were built. Incredibly, in 1914, Ford had built more cars than all the other manufacturers combined.

THE SATELLITE

The race to dominate orbital space

Nominated by **Professor John Zarnecki**, professor of space science, Planetary & Space Sciences Research Institute, The Open University

'Extra-terrestrial Relays' was the title of a paper by the legendary science fiction author and futurist Arthur C. Clarke. The idea behind the paper, published in *Wireless World* magazine in 1945, was that geostationary satellites would orbit the Earth and act as telecommunication links. Clarke envisioned combining the technologies in rockets, wireless communications systems and radar to put this into practice. Hence, Clarke has been credited with the invention of the satellite, but some claim the idea was already in the general consciousness at that time. Regardless, 12 years later, the first artificial satellite made it into orbit and Clarke's vision became a reality.

Sputnik 1 was simply a hollow steel ball, just 58 cm across, with a radio transmitter inside, but when its radio started beeping as it circled hundreds of kilometres above the Earth on 4 October 1957, it heralded a new age for mankind – the space age. The satellite was a much smaller device than originally intended. Tangled bureaucracy had hindered the development of the Soviet Union's preferred model – Object D. Rushing to beat the Americans to launching a satellite, Sergei Korolev, rocket genius and mastermind of the Soviet space programme, went against his bosses wishes and created the much smaller and simpler *Sputnik 1*.

The first launch attempt, on 15 May 1957, failed 100 seconds into the flight because of a fire in one of the side-rockets. The Russians tried again numerous times and failed. Finally, on 4 October 1957, lift-off went according to plan and the first full Earth orbit was seemingly snatched from under the noses of the Americans.

Flopnik

The Americans were in fact blissfully unaware of the Soviet Union's impending triumph. When they discovered that *Sputnik 1* was circling over their heads that evening, the CIA was devastated that their intelligence hadn't picked up on the programme.

Over the next couple of months, America scrambled to catch up. Their plan was to stow away a grapefruit-sized satellite in their *Vanguard TV3* rocket but, just two seconds after launch on 6 December, the rocket exploded in flames. Ironically, the satellite escaped the fireball and started beeping its signal to any radio tuned in to listen.

A red-faced America felt drastic action was needed to remedy their 'Flopnik'. Members of the US Satellite Committee had recommended a few weeks earlier that a special space agency was needed, and eight months later NASA was formed.

Rocket Dog

After the success of *Sputnik 1*, Soviet Premier Nikita Khrushchev immediately called Korolev to congratulate him and to set him another challenge: to put something else into orbit. Korolev already had the idea for *Sputnik 2* up his sleeve and the design was virtually ready to go. On 3 November 1957 – over a month before the US had even made its *Vanguard TV3* attempt – *Sputnik 2* took to the skies. Larger than its predecesor, the satellite could be seen by the naked eye, reflecting sunlight as it streaked across the sky. What made this satellite special was its cargo.

On board was Laika (meaning 'Barker' in English). This light-coloured female dog sat in a pressurised module beneath a sphere containing the radio transmitter. Her light colour made it easier for mission control to see her via the on-board cameras, and the dog was female so the nappies she wore could be simpler in design.

As she hit Earth orbit, the mongrel that had been found wandering the streets became a national icon of the space age. Although the module had been equipped with poisoned food which would euthanise Laika after ten days, 45 years passed before the public found out that she had died just six hours into the flight from stress and overheating when the protective insulation was unintentionally torn away and the thermal control system failed. One of the team, Oleg Gazenko, later reflected: "The more time passes, the more I am sorry about it. We did not learn enough about the mission to justify the death of the dog."

America decided satellites should be used to carry out some truly useful science. On 31 January 1958, the US finally launched the satellite *Explorer 1*. On board was a temperature sensor, particle detector and a microphone to pick up the sounds of any meteorites colliding with the satellite. The scientists from the Jet Propulsion Lab, William Pickering, Wernher von Braun and James Van Allen, were rapturously applauded at the press conference the next morning.

During the 1960s, satellites were designed for all sorts of tasks. Project Echo, which began in May 1960, tested the deflection of radio signals for communication. The *Discoverer* satellites, the first of which was launched on 28 February 1959, were like 'eyes in the sky', replacing the need for spy planes.

This was the dawn of a new age of communication and surveillance. The space age had developed into the space race.

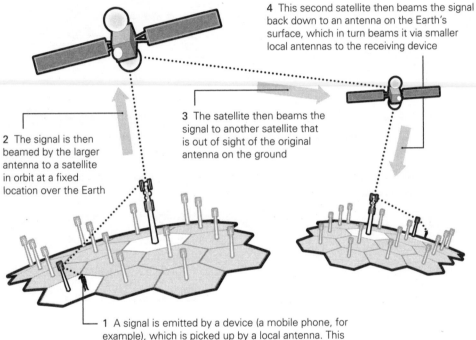

4 This second satellite then beams the signal back down to an antenna on the Earth's surface, which in turn beams it via smaller local antennas to the receiving device

3 The satellite then beams the signal to another satellite that is out of sight of the original antenna on the ground

2 The signal is then beamed by the larger antenna to a satellite in orbit at a fixed location over the Earth

1 A signal is emitted by a device (a mobile phone, for example), which is picked up by a local antenna. This antenna in turn forwards the signal on to a larger antenna

THE V-2 ROCKET

The weapon that changed the course of history

Nominated by **Graham Southorn**,
editor of *Sky at Night* magazine

In the late 1920s, a debate was raging between physicists at VfR, the amateur German Society for Space Travel. The German military had offered funding for those who were prepared to work on developing rocket technology for military purposes. Some wanted nothing to do with the military; others saw it as a necessary evil in order to get some support for their work. Wernher von Braun and World War I veteran Walter Dornberger were two scientists prepared to take the gamble.

"The V-2 rocket changed the world more than anything else in history, as it kick-started a global chain of events."
Graham Southorn

Through the 1930s, Dornberger was in charge of developing the so-called 'A-series' of rockets. The A1 never made it off the drawing board. The A2 incorporated a key new piece of technology (a spinning gyroscope, which stabilised the rocket), while the A3 was a more powerful model. But it was the A4 that was the big breakthrough - a turbo-charged rocket with advanced fuel injection.

Fortunately for the Allied forces, some funding for the A4 was redirected into support for other military projects and the development of the A4 was delayed. Mass production of the rocket started in the early 1940s, but a combination of the rapidly advancing Soviet Army and a devastating raid by Allied forces on the rocket manufacturing base at Peenemünde in 1943 put a big spanner in the works for the Germans. The raid killed one of their key engineers Walter Thiel who had designed the A4's super-engine. A new base and a new name for the rocket were needed.

Subterranean Lab

The programme relocated to an underground base in the Mittelwerk network of tunnels under Kohnstein Mountain in central Germany, a location fit for a Bond film set. Alongside other weapons, the A4 was honed by its creator, von Braun, under its new name *Vergeltungswaffe-2*, meaning the 'Reprisal Weapon 2', or V-2. The infamous German minister for propaganda, Joseph Goebbels, hoped this name would strike fear into the Allied forces.

The V-2 certainly was a ferocious weapon. Unlike its noisy predecesor (V-1), it made a silent approach, the engines shutting off after it hit the peak of its trajectory, gliding quietly in on its target.

But the V-2's heyday was too late for the Germans – they were already rapidly retreating across Europe. "The V-2 wreaked terrible destruction on London, but arguably cost Germany the war, because it spent so much money on V-2 development but it wasn't ready in time to turn the tide against the Allies," says Southorn.

Jumping Ship

Quick to spot that Germany was a sinking ship and the US the best patron, going against German orders, von Braun and some of his team packed up the missile blueprints and stashed them in a disused mine near Mittelwerk in the dead of night, safe from the eyes of the invading Soviet army.

Fortunately for von Braun, the Americans were on the lookout for him, keen to capture the brains behind the V-2, realising the enormous potential of the rocket. On 12 September 1944, seven German scientists signed up to Operation Overcast, which later became known as 'Paperclip'. They agreed to work for six months in the US, but had to leave their families behind. Von Braun was one of them.

Other key German scientists decided to turn east to the Soviet Union. And so, with the rocket expertise balanced fairly equally between East and West, the next stage of the missile race began.

The Cold War

After the Soviets had squirrelled away some of the remnants of the V-2 technology, they set about developing it. "As the Soviet Union could no longer afford conventional nuclear bombers in the form of planes, they began to develop a new nuclear warhead delivery mechanism which made warfare cheaper," says Southorn. "This led to the nuclear arms race and, indirectly, to the Cuban missile crisis."

The Soviet Union was also intent on using the technology to venture into space. Soviet rocket genius Sergei Korolev was the brains behind the launch of *Sputnik 1*, the world's first satellite (see page 349), as well as the rocket that launched the first man, Yuri Gagarin, into space.

Across the Atlantic, controversy over the existence of the German scientists in the US continued to rumble on, not only because they had been flown in with no visas, but also because of their former close links with Nazi Germany. Von Braun was no exception. He had joined the Nazi party in 1937 and had been promoted to the position of SS lieutenant. He claimed it was the only path open to him if he wanted to continue his aim of creating a rocket fit for space travel.

Von Braun got off lightly after the war (Dornberger, who had been more entrenched in the Nazi Party, was imprisoned for two years, while others were sentenced to death). However, Von Braun's influence on rocket technology was immense, and, despite the murmurings of dissent, he was considered a worthwhile investment for the US. He was to be one of the key figures behind the *Saturn V* rocket that powered NASA into space and to the Moon.

"V-2 technology not only changed the face of warfare and space exploration, but also led to numerous other important discoveries," says Southorn. "The first Moon landing resulted in all sorts of inventions, such as cushioned soles for running shoes, cordless power tools and lightweight firefighting equipment. And, arguably, it led to the internet as we now know it. The military quickly realised that its forerunner, a distributed computing network called 'ARPANET', could be developed into a command structure capable of surviving a nuclear strike."

THE REUSABLE SPACECRAFT

The race to build a craft capable of going into space and returning in one piece

Nominated by **Elon Musk**, entrepreneur and co-founder of PayPal, Tesla Motors and SpaceX

"The one invention I think will really change the world is a truly reusable rocket as it's fundamentally required for life to become multi-planetary."

Elon Musk

Yuri Gagarin was nervous – and rightly so. Previous spaceflight guinea pigs included dogs and monkeys. Never before had a human gone into space. On 12 April 1961, at Baikonur Cosmodrome in Kazakhstan, Gagarin buckled up and jetted off for a trip to outer space. As he orbited the planet, he marvelled at its hazy blue band of atmosphere, humming the words of a Russian song: "The Motherland hears, the Motherland knows where her son flies in the sky."

In the 50 years since Gagarin's trip, Man has stepped foot on the Moon, sent probes to the outer reaches of the Solar System, landed rovers on Mars, and built the International Space Station (ISS). Much has changed in space technology, often driven by harsh lessons learnt through terrible accidents like the *Challenger* disaster.

One of the key developments has been the creation of semi-reusable rockets, of which Gagarin was a key instigator. During his later life at the Russian cosmonaut facility at Star City, he researched some of the first designs for such a craft.

Thunderbirds Aren't Go

The dream in the 1950s was a *Thunderbird 1*-style vertical take-off and landing rocket. The reality was that construction materials and engine technology weren't even ready for a reusable rocket, let alone one that landed vertically. But as material scientists developed tougher substances and engine technology advanced, engineers including Philip Bono began suggesting innovative designs. It wasn't until designs for the space shuttle began in the late 1960s, though, that even a partially reusable rocket became a possibility.

The shuttle first took to the skies on 12 April 1981 – exactly 20 years after Gagarin made his epic first trip. It has just retired after three decades of launches, carrying and repairing satellites, ferrying astronauts to the Mir space station and transporting more than 200 people from over 15 countries to the ISS. The *Discovery* shuttle alone made 39 trips, spent 365 days in orbit, circled the globe 5,830 times and travelled over 237 million km – equivalent to over 308 round trips to the Moon or more than one and a half times to the Sun and back.

There were three main sections to the Space Shuttle: the orbiter, solid rocket boosters and the external tank. The orbiter carried seven astronauts and one very expensive toilet (reportedly costing millions). Each heat-proof tile on the exterior of the orbiter cost as much as $1,000, but each was vital to stop the aluminium frame melting on re-entry through the Earth's atmosphere, when temperatures reached over 1,650°C.

The solid rocket boosters contained fuel made of powdered aluminium mixed with an oxidiser, a catalyst and a binder, together with synthetic rubber that was cured into massive blocks. The external tank carried 1.5 million pounds of propellant which helped to get the shuttle into space and provided structural support during the launch, absorbing 7,800,000 pounds of thrust. It was jettisoned eight and a half minutes into flight at an altitude of around 113 km.

The external fuel tank was the only non-reusable part. The engines and boosters were reused, but only after a few months of refitting work. Wherever the shuttle landed around the world, it was ferried on the back of a specially converted Boeing 747 to the Kennedy Space Center in Florida.

Dragon Capsule

With the shuttle's limitations in mind, for the last few decades, governments have been keen to build a fully reusable spacecraft. Ronald Reagan envisaged a scramjet plane as a possible 'single-stage-to-orbit' craft, while British engineers designed the hydrogen-powered craft *Skylon* that would take off from a runway. Sadly, extortionate costs grounded the programme.

A new fleet of spacecraft is currently being developed by private companies, such as Sierra Nevada Corp's Dream Chaser, which looks a bit like a mini-shuttle. Another is the *Dragon* capsule created by Musk's SpaceX. Though it will initially only be used for ferrying cargo to the ISS, it's been designed to meet the more stringent requirements for flying astronauts. As we bid the shuttle farewell, we welcome in a new age of reusable spacecraft.

The World of Technology

THE CAMERA

How photography captured society

Nominated by **Dr Patricia Fara**, senior tutor
at Clare College, Cambridge, whose latest book
is *Science: A Four Thousand Year History*

Around 2,000 pictures are uploaded to Facebook every second. Today, it's so simple and quick to take a photo upload it to a social networking site. But it hasn't always been that way.

The first ever photograph needed an exposure time of eight hours. It was taken by Joseph Nicéphore Niépce in 1822, but the image was later lost when he tried to replicate it. A few years later, working alongside fellow Frenchman Louis Daguerre, Niépce developed a technique he called 'heliography'. By exposing a pewter plate coated

"Communicating visual information is essential for scientific progress, and a good picture can be worth a thousand words. Drawings are not always reliable, but the invention of photography enabled scientists to record objective, accurate images, even of formerly invisible phenomena, and transmit them around the globe."
Patricia Fara

with a type of tar that was sensitive to daylight, and washing it with lavender and petroleum, then exposing it to iodine, he captured the first permanent photograph.

As a renowned theatre designer, Daguerre needed a quicker way to take a picture to record his designs. In 1833, he found it – creating an image in less time by treating an iodine-coated silver plate with mercury fumes before fixing it with a dowsing in salty water. In1838, he created the first image of a person when he captured a pedestrian stopping for a shoe shine. Coy about his invention, Daguerre initially refused to let slip anything about the technique, but, in return for a healthy pension, he eventually revealed all. Taking a photo was still a lengthy and expensive process, though.

A simpler and cheaper method for fixing an image was needed. Chemist and astronomer John Herschel (whose father, William, discovered the planet Uranus and aunt, Caroline, was also a notable astronomer) found that iron salts produced a fixed blue image. This 'cyanotype' method was the predecessor to the modern blueprint process but, crucially, was inexpensive and easy to do, making it a commercial success. In 1840, William Fox Talbot developed the calotype method. This was the first time anyone had used the negative/positive process that is still used today, and it was revolutionary in that it allowed the image to be reproduced numerous times.

The advent of film was the next major breakthrough in photography. Flexible nitrate-based plastic film was created by George Eastman. The film was rolled up and loaded into relatively cheap cameras. Once you'd taken around a hundred pictures, you took your camera along to a developer at Kodak for prints to be made. Colour film was first produced in 1935.

Few people, apart from professional photographers, still work with film. The ease of snapping a digital shot has revolutionised photography.

THE CAMERA OBSCURA

Chinese philosopher Mozi was the first to note that when light rays pass through a pinhole into a dark room they created an inverted image – an early form of a camera obscura. This was way back in the fifth century BC. Through the ages other renowned scientists like Aristotle and Alhazen played around with the effect, and in the thirteenth century the English natural philosopher Roger Bacon described how the camera obscura was a safe way to view a solar eclipse.

Camera obscuras were created in darkened rooms or tents, but following the work of Robert Boyle and Robert Hooke, portable models had been invented by the eighteenth century – and the earliest cameras were born.

Digital image photography didn't develop until the 1950s, and was used for tasks such as mapping the Moon. The Nikon F3 was the first digital camera to hit the market in 1991.

LIQUID CRYSTAL DISPLAY

How a carrot paved the way for the discovery of LCDs

Nominated by **Rahiel Nasir**, technology commentator and writer

Liquid crystal displays (LCDs) can trace their origins back to the humble carrot. In 1888, after extracting cholesterol from a carrot, Austrian botanist and chemist Friedrich Reinitzer noticed that the cholesterol had a strange property: it was neither liquid nor had solid crystals, but was somewhere in between. Reinitzer found that the cholesterol bizarrely had two melting points and reflected light in interesting ways.

"LCDs have become an integral part of everyday life. They are in laptops, mobile phones, flat screen TVs, GPS devices, as well as a whole host of commercial and professional equipment."
Rahiel Nasir

While the discovery generated a bit of a buzz among scientists, at the time no-one could find a useful application. Through the early 1900s, various scientists played around with the liquid crystals, trapping them

HOW LCD TECHNOLOGY WORKS

LCD uses layers of liquid crystals, polarising film and electrodes to produce a picture. The polarising film orientates the light in one direction. When there's no current from the power source to the electrodes, the light entering the front of the LCD simply hits the mirror and bounces back out. When a current is switched on, however, some of the liquid crystals 'untwist' from their naturally twisted state, and the next polarising film stops the light passing through – so that pixel remains dark. This combination of twisted and untwisted crystals allows different levels of light to bounce back off of the mirror, forming the images that we see on the screen.

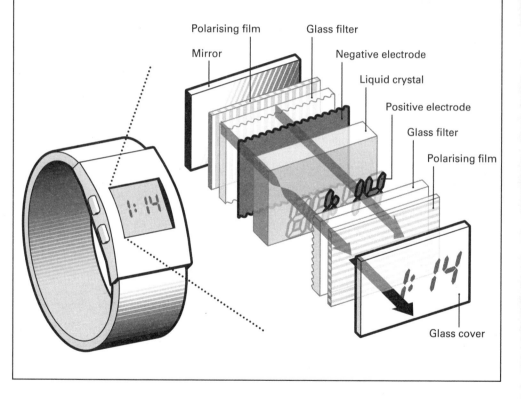

Polarising film

Mirror

Glass filter

Negative electrode

Liquid crystal

Positive electrode

Glass filter

Polarising film

Glass cover

between thin layers of plates. But it took a large corporation to spot the first practical application.

Realising its potential, the Marconi Wireless Telegraph Company patented the 'liquid crystal light valve' in 1936. The real breakthrough, however, came in the sixties when engineer George Heilmeier realised the significance of work by Richard Williams, a colleague at the Radio Corporation of America. Williams had discovered that passing an electric current through a thin layer of liquid crystal created a striped pattern. Heilmeier immediately saw its massive potential and suggested using liquid crystals in displays. The LCD was born.

By the seventies, LCDs were inside wristwatches, and, by the nineties, the technology was being used in TVs, eventually replacing cathode ray tube displays. Today, LCDs are integral to most of our gadgets, from clocks to laptops to DVD players to smartphones. As Nasir says: "LCD screens are everywhere in the modern world."

FIBRE OPTICS

The invention that will spark the next communication revolution

Nominated by **Michail Bletsas**,
director of computing, MIT Media Lab

Fibre optics have revolutionised many industries, but the idea behind them isn't that new. Way back in the 1840s, a number of physicists demonstrated how light could be transmitted down a tube. By the early twentieth century, dentists had found a use for this phenomenon to light up the mouths of patients. In the 1950s, Basil Hirschowitz, a physician at the University of Michigan, came up with the idea of using fibre optics to examine the stomach and intestines of his

"With their ever-increasing capacity and their enablement of the information economy, fibre optics are the nerves of our increasingly globalised society. As we push past the current speed record of 10 terabits per second (that is, 10 trillion bits per second), their ability to pull everybody closer will greatly exceed the centrifugal effects of our societal differences."
Michail Bletsas

patients. This fibre-optic endoscope was a vast improvement on the rigid metal pipe used before.

A fibre-optic tube contains a long glass strand. It works a bit like shining torchlight down a bending hallway that's covered with mirrors: the light bounces to the other end of the hallway by reflecting off the mirrors. The high-quality glass contains far fewer impurities than a pane of glass. The impurities in a pane of glass mean the thicker it is, the poorer the view, but if a deep ocean was made of the pure glass used in fibre optics, you'd be able to clearly see the seabed from the ocean's surface.

In fibre-optic cables used in telecommunications to transmit digital data, the glass core, which is as thin as a human hair, is surrounded by a layer of cladding that reflects any stray light back into the core, and an external buffer layer protects the fibre from moisture and damage.

Most broadband networks are ADSL, which uses copper phone lines, so transfer speeds are relatively slow. Even the latest ADSL2+ only reaches download speeds of about 24 megabits per second. A network using fibre-optic cables can go up to 100 megabits per second.

Fibre-optic cables are the nerves of CERN's Grid computing network. While the worldwide web simply shares information on computers, the Grid also shares computing power and storage capacity. So scientists can log on to the Grid from their computer, and PCs all around the world will carry out the work for them. Without fibre-optic cables, complex calculations couldn't be conducted quickly. This system has come into its own by processing the vast amounts of data produced by the Large Hadron Collider, some 15 petabytes (that is, 15,000 trillion bytes) of information a year, which works out at almost 1 per cent of all the digital information produced annually on the planet.

Once fibre optics move into our homes, communication speeds will be 10,000 times faster than today's connections, allowing us to truly instant message one another or Skype with holographic images. A communication and entertainment revolution is just around the corner.

THE TELEGRAPH

The invention that
led to Morse Code

Nominated by **Luis Villazon**, *Focus* contributor,
whose works include *How Cows Reach the Ground*

Thousands of years ago, long-distance communication used to involve sending signals by beating drums and fanning smoke from fires. Then came the age of writing and letters carried by people – or pigeons. It wasn't until 1799 when Italian scientist Alessandro Volta discovered how to generate a steady current that an electrical means of communication became a reality.

In 1830, American scientist Joseph Henry succeeded in sending an electric current down a 1.6 km-long wire to an electromagnet, which struck a bell. A couple of years later, the story goes that, following the death of his wife, Samuel Morse returned to Europe to further his artistic career. Aboard the ship he got chatting to some scientists who inspired him to investigate the telegraph. Realising its massive potential, Morse ditched his career as a painter and, alongside engineer Alfred Vail, set about creating a better version. It was through this device that they devised a new telegraphic alphabet.

Letters were represented by a series of short and long electrical signals, the dots and dashes of the now famous Morse code. Such a simple idea, yet so effective.

Through the 1840s telegraph lines sprung up all over America and Europe. In 1839, a line was installed along the Great Western Railway, which proved vital in catching a murderer on the run in 1845, as a telegram tip-off meant that police were waiting for him when he arrived at Paddington Station.

The first line to be laid across the Atlantic prompted Queen Victoria to send the world's first cross-Atlantic telegram in 1858 to US President James Buchanan, with the congratulatory message: "Glory to God in the highest; on earth, peace and good will toward men." The line was destroyed a month later when English surgeon and chief electrician for the Atlantic Telegraph Company, the aptly named Wildman Whitehouse, managed to apply a surge of voltage which fried the wire. A lasting connection was eventually installed in 1866 by Isambard Kingdom Brunel's ship, the SS *Great Eastern*.

"When the telegraph killed off the pony express, it gave birth to the career of journalism. Instead of waiting for local editions of newspapers to be physically carried to another city or country, important stories could be sent instantly around the world and reported simultaneously in every major city. By finally cleaving communication from travel, the telegraph was the first technology to turn information into a commodity. It was the Victorian internet."

Luis Villazon

For many decades the telegraph was the ultimate method for long-distance communication, even being used in the twenty-first century, with the last telegram sent in 2006. In the 1870s, another invention stole the limelight – namely the telephone.

DID YOU KNOW?

The word 'telegraph' is derived from Greek and means 'to write far' – a literal description of the purpose of a telegram.

THE TELEPHONE

How Alexander Graham Bell stole the limelight for the invention that transformed communication

Nominated by **Lesley Gavin**, futurologist and adviser to the European Commission

As Italian Antonio Meucci electrocuted one of his patients while 'treating' him for rheumatism, he thought he heard the patient's yelp from an electrical conductor near his own ear. After further investigation, the Italian found that the sound had passed down a copper wire, attached to the electrical conductor, which then vibrated to produce the yelping sound.

In the months that followed, he set about creating a device that picked up the sound vibrations on

"The invention of the telephone, and the ability to communicate instantaneously over long distances, changed the way we thought about communications. Not only did it change the way people did business, but it also changed how we socialised. It paved the way for fax machines, email, and ultimately information sharing services such as YouTube and Facebook."
Lesley Gavin

a membrane. This vibrating membrane caused an electromagnet to move, which in turn affected a current in a wire. The reverse of this process reproduced the sound at the other end of the device. And a proto-telephone was born.

So how is it that the Scottish scientist Alexander Graham Bell is now regarded as the inventor of the telephone?

Meucci got the first patent for his device in 1871, but in 1874 a lack of cash meant he couldn't afford the $250 needed to secure the patent. Bell was waiting in the wings. He shared a lab with Meucci and realised that here was his chance to develop the Italian's proto-telephone into a commercial success.

Bell had funding to develop a system for sending numerous telegraph messages using different signal frequencies. In his telephone device, a thin membrane vibrated when sound was directed at it. This moved an iron rod in front of an electromagnet, to produce an electric current. At the other end, the reverse process occurred: the electric current caused the iron rod to move, which in turn moved the membrane, to produce sound waves.

Bell patented this device on 14 February 1876, reportedly beating Elisha Gray to the patent office by a matter of just a few hours. No love

DID YOU KNOW?

- In 1936, London telephone operator Miss Jane Cain became the first voice of the speaking clock.

- In 1937, the emergency number 999 was first introduced to London.

- In 2008 alone, Ofcom allocated around 126 million numbers.

INSIDE A MODERN TELEPHONE

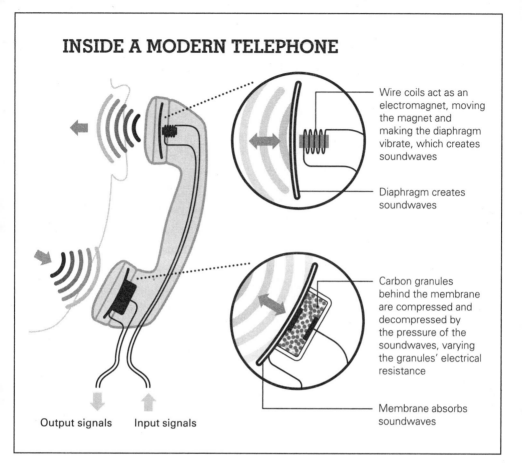

Wire coils act as an electromagnet, moving the magnet and making the diaphragm vibrate, which creates soundwaves

Diaphragm creates soundwaves

Carbon granules behind the membrane are compressed and decompressed by the pressure of the soundwaves, varying the granules' electrical resistance

Membrane absorbs soundwaves

Output signals   Input signals

was lost between the two men. Gray accused Bell of stealing his idea and a patent officer claimed he'd been bribed to give Bell the patent. Bell fought off over 600 lawsuits throughout his life and won every case. Even Antonio Meucci filed a lawsuit against Bell and came close to victory when he died in 1889. However, the unfortunate Italian was finally recognised by the US Congress in 2002 as the true inventor of the telephone.

THE MOBILE PHONE

How radio waves sparked a
telecommunications revolution that
led to the smartest of smartphones

Nominated by **Spencer Kelly**,
presenter of BBC technology show *Click*

*"They say knowledge is power.
Access to information doesn't just
enhance lives, it changes society.
Mobile telecommunications can
bring the democratising power of
information, and the processing
power of the internet, to everyone,
everywhere: whether it's using
a mobile phone to translate your
words from one language to another
as you speak them, or educating
children in developing countries."*
Spencer Kelly

The advent of radio waves and wireless transmission ignited another telecommunications revolution in the twentieth century. Being able to broadcast a single signal wirelessly to a large number of receivers was the key that unlocked the mobile revolution.

The first mobile phones weren't really mobile at all: they just didn't use the mains electricity supply. Early models, such as the Motorola

4500x, were handsets hooked up to large, heavy batteries. But the battery didn't last long, allowing just 20 minutes of talk time. With a price tag of over £1,000, it is easy to see why people thought they would never catch on.

Former Motorola vice-president Martin Cooper is credited with inventing the mobile phone, and his team created the first hand-held cellular phone. Cooper made the first call on 3 April 1973, as he strode into a hotel lobby for a press conference. He joked that the gadget was inspired by the communicator used by Captain Kirk in *Star Trek*.

In 1987, the Motorola 8000X series was launched in the UK. At £1,200, it became the must-have gadget for the rich – and also for Del Boy in TV sitcom *Only Fools and Horses*. Then, in the nineties, came the digital revolution, with the switch from analogue to digital technology. Combined with new operators, including Orange and One2One, the digital networks drove down the price of mobile phones.

The first SIM (subscriber identity module) card was created in 1991 by Munich smart-card maker Giesecke & Devrient. It contains a whole host of information on everything from its unique serial number to security authentication and ciphering information. It enables users to switch phones while keeping the same number by moving the card from one handset to another.

As SIM cards were adopted commercially, manufacturers continued to battle it out, with Motorola's clamshell-style 88 g mobile that had 4 hours talk time, 47 hours standby and a vibrating ringer, up against the likes of Nokia's slide-opening mobile that starred in the 1999 film *The Matrix*.

In the same year triband technology was launched, making mobile phones capable of operating on three different frequencies, so they worked in both Europe and the US – useful for globetrotting businessmen. In

DID YOU KNOW?

- One-third of the global population make calls using their mobile phones.

- There are currently around 3 billion mobile phones in use worldwide.

- The number of mobile phones in Britain now exceeds the number of people.

- About 90 million mobile phones are lying unused in homes throughout the UK, weighing in at 11,250 tons, five times the weight of the London Eye.

- SIM cards come mounted in a credit card-sized 'carrier' because some older phones actually used this full-size SIM. The much smaller card with the corner cut off is technically a mini-SIM.

1994, Ericsson created a novel way of wirelessly transferring data between compatible devices over short distances, known as 'Bluetooth'. Then came the advent of 3G (third generation) technology, with increased bandwidth and more diverse applications, or 'apps'.

Downloading Android or Apple apps onto your smartphone gives you access to anything from the latest news of exoplanet discoveries to the massive hit *Angry Birds*. At the start of 2010, more than 3 billion apps had been downloaded by iPhone users worldwide. As Apple says: 'There's an app for everything.'

Today, our mobile phones are not just for making calls. Blackberries and other smartphone handsets let us work, rest and play: we can use them to check emails remotely, snap photos and listen to podcasts. Smartphones are even starting to replace the need for hard drives in computers. Some models already have not one but two processors. Their vast computing power means they can be plugged into a dock at the back of a monitor or TV screen, transforming it into a web-browsing computer. With the superfast mobile 4G network reportedly due to roll out in 2012, you could surf the net at broadband speed or instantly load iPlayer programmes on your telly just using your smartphone's network.

THE TRANSISTOR

Transforming our technological world

Nominated by **Jimmy Wales**,
co-founder of Wikipedia

Just as cells are the building blocks of life, transistors are the building blocks of modern technology. They are integral to almost every bit of technology in our lives, from computers to cars.

"The invention of the transistor changed everything. It made possible the creation of computers and, indeed, modern electronics of all kinds. This led to the internet and advances in all areas of science and technology."
Jimmy Wales

Before transistors existed, devices called 'vacuum tubes' were used in electrical circuits in everything from TVs to telephones. They were overly large and unreliable, and, as they had to heat up before they could work, they weren't very energy efficient. Amazingly, an alternative had existed for two decades.

In the 1920s, physicist Julius Lilienfeld filed patents for a device with three electrodes – electrical conductors that are used to make contact

MOORE'S LAW

In 1965, Intel co-founder Gordon Moore observed that the number of transistors on microchips doubles approximately every two years. This statistic is now called 'Moore's law'.

Moore's law has proved to be uncannily accurate. Indeed, it has held true for half a century already. If the diving price of chips is plotted on a graph against chip processing power, the line is virtually straight for 50 years. Moreover, the line extends back 100 years if the vacuum tube device and, before that, the mechanical hand-crank machines are taken into account.

But Moore's law will finally come to a halt, possibly as soon as 2020, when transistors on wafer-thin silicon chips are the size of just one atom.

with non-metallic parts of circuits. No-one knows if Lilienfeld actually invented the device, but it was two decades before someone spotted how useful it could be.

In the 1940s, someone at Bell Labs in the USA cottoned on to Lilienfeld's three-electrode device, and tasked the brightest of its staff to find a practical use for it. By 1947, the physicists John Bardeen and Walter Brattain had created an amplifying circuit, where the output power was larger than the input when electrical contacts were applied to a crystal of germanium. William Shockley then improved the device, building what's now known as a 'junction transistor'.

HOW TRANSISTORS WORK

In simple terms, transistors are essentially on/off switches. They control electron movement. So, rather like turning a tap on or off, or up to full power, transistors can both amplify and switch electronic signals, meaning the voltage or current in a circuit board can be controlled precisely.

Each transistor has at least three terminals that connect to other circuits. A voltage or current applied to one pair of the transistor's terminals changes the current flowing through another pair of terminals.

The team from Bell Labs was awarded a Nobel Prize in physics in 1956, but its invention would not have been possible without a breakthrough in refining elements and something known as 'doping'. Doping occurs when impurities are added to elements to make them worse insulators. These so-called 'semiconductors' have both insulating and conducting properties, and so their conductivity can be varied, which is essential for transistors.

The hard grey-white element germanium is normally an insulator, but when impurities are added, it becomes a good semiconductor. After World War II, advances in doping and germanium refinement meant that germanium could be made into a perfect semiconductor for transistors. Germanium transistors were used worldwide for over 20 years, until a new element was found to work even better – and was cheaper!

Silicon Valley

In 1954, George Teal at Texas Instruments created the first transistor made from silicon – an element in the same group in the periodic table as germanium. Silicon, in the form of silicon dioxide, is the second most abundant element in the Earth's crust, and hence cheaper to extract than germanium; in addition, it made more reliable transistors.

Silicon transistors revolutionised computer design and it wasn't long before microchips were being crammed with millions of transistors, giving them the ability to perform ever-more complex calculations (see page 386). The boom in computer engineering had begun and the region in California where much of this activity was based became known as Silicon Valley.

DID YOU KNOW?

- The word 'transistor' is a portmanteau of 'transfer' and 'resistor'.

- Transistors now exist that are the width of a single strand of hair.

- One of the first pre-transistor computers was the Electronic Numerical Integrator and Computer (ENIAC), which contained 17,000 vacuum tubes and weighed 30 tonnes.

THE MICROCHIP

The tiny device that led to the dawn of
PCs and heralds a new era of plastic chips
and quantum computers

Nominated by **Lord Robert Winston**, professor of
science and society at Imperial College London

Micro in name and nature, but the impact of microchips on society
is anything but minuscule. Made up of various electrical components
attached to the same silicon chip, they are used in everything from
mobile phones to personal computers to credit cards.

Before microchips, circuit boards were printed and then components
individually wired and soldered together. But in the late 1950s, an
American electrical engineer called Jack Kilby had a brainwave.

Kilby spotted that, if each component was made from the same
semiconducting material, they wouldn't need to be wired together. In
1958, he demonstrated the first working chip made from the element
germanium to colleagues at the engineering firm Texas Instruments.
Not only did making all the components from the same element keep
the costs down, it also cut the time taken to build them.

Within a decade, engineers were cramming hundreds of components onto the same chip. Far higher densities of components became possible with the advent of photolithography, where a combination of chemicals, gas and high-frequency light is used to build up components on wafer-thin slices of silicon. The resulting 'microchips' contained huge numbers of tiny electronic components able to carry out ever-more complex computations, and they found their way into everything from microwave ovens to mobile phones to laptops to supercomputers.

By 2008, microchip-maker Intel announced that it had succeeded in packing over 2 billion components onto one tiny chip. But such achievements are pushing current technology to its limits. As the density of components goes up, it becomes ever harder to prevent problems of interference between them. More computing power also means higher power consumption and also more problems in dealing with the heat generated.

Chip Future

To get round such problems, engineers are turning to new materials. While we've used graphite for a long time in everyday objects, such as the humble pencil, the atom-thick sheet of graphite known as 'graphene' was only discovered in 2004. By 2009, Massachussetts Institute of Technology had developed an experimental super-fast graphene chip, which could allow conventional silicon chips to cope with even higher computation speeds.

Quantum physics – the laws of the subatomic world – will also play an increasing role in chip technology. Some experts believe that future chips will exploit the bizarre abilities of quantum systems, such as the ability of particles to have different properties at the same time. This would open the way to ultra-fast parallel processing computers able to solve a huge number of problems simultaneously, bringing a colossal increase in speed. There are huge challenges, but researchers around the world are working to solve them. In 2011, a team from the University of California announced the creation of a quantum chip that they believe could be scaled up to solve real-life problems.

In the meantime, Jan Genoe and his colleagues at IMEC, a semiconductor research centre in Leuven, Belgium, have created a chip made from plastic, polyethylene naphthalate, similar to the plastic used to make bottles and wrap sandwiches. It can only run at a speed of 6 Hz, which is about one-millionth of the speed of the average laptop. Silicon chips need a billion-dollar plant to etch on the tiny electronic components, so the hope is that plastic chips will be cheaper to make – and crucially enable the development of flexible display screens.

It seems that the revolution that Jack Kilby began half a century ago may be about to start all over again.

DID YOU KNOW?

The world's smallest microchip, launched in 2006 by Hitachi, is the size of a single speck of dust and can be used in gift coupons and parcel wrappings.

THE COMPUTER

From the Analytical Engine to the iPad

Nominated by **Professor David Deutsch**, visiting professor of physics at the Clarendon Laboratory, Oxford University, and author of **The Beginning of Infinity**

The word 'computer' was first used in 1613 to refer to someone who carried out calculations. It was still being used in that way even in the 1940s – despite a brilliant English mathematician coming up with the idea in the 1820s of a machine that could be programmed to perform calculations automatically.

Charles Babbage was born in London and studied mathematics at Cambridge University, eventually being elected Lucasian Professor of Mathematics at the university in

"Charles Babbage wanted to produce more reliable navigational tables. A century later, Alan Turing addressed a problem in mathematical proof theory. Both realised that their idea for solving those parochial problems could also be used to build machines with the unprecedented ability to compute anything computable. Result: the computer revolution."
David Deutsch

1828 – the same position that was held previously by Isaac Newton and in modern times by Stephen Hawking. Inspired by the devices born of the Industrial Revolution, Babbage dreamt up the idea of a so-called 'Difference Engine' that would be able to evaluate any useful sets of numbers.

After Babbage presented the idea to the Royal Astronomical Society, the government granted him funding to build the machine, but arguments broke out between engineers over cash. Unable to secure funding to construct the colossal device, Babbage had to settle for building just part of it for demonstration purposes. This is now recognised as the first automatic calculator. Unphased, Babbage improved the machine's design to produce the Difference Engine No. 2 and went on to design his so-called 'Analytical Engine'.

Using the idea of punched cards to program patterns on looms from the textile industry, Babbage's Analytical Engine had cards that would have made it the first programmable computer. But this machine was also never built. Nevertheless, Ada Lovelace, the daughter of English poet Lord Byron, wrote programs for the cards – and is now considered the first computer programmer.

It was only after the first electromechanical computers were built in the twentieth century that their inventors realised how Babbage had anticipated so much in computer design. He was way ahead of his time.

Turing's machine

As with so many inventions, war accelerates developments. World War II did just that with the evolution of the computer, as demand was high for machines that could crack complex codes. So the powers that be were on the hunt for the brightest minds who were already working in this field.

Born into a middle-class family in London in 1912, Alan Turing followed in Babbage's footsteps, studying the challenging mathematical tripos course at Cambridge University. He moved on to a fellowship at King's College, where he worked on probability theory. It was later, while at Princeton University in the US, that he published a paper in 1936 that proposed the idea of a computing machine that uses a long strip of tape as its memory. This was a theoretical device and the basic idea involved a tape divided into cells, each cell sporting a single symbol encoding some information. As the tape was 'read', these symbols would instruct the machine what to do next. Provided that there was a clear algorithm, any mathematical problem could be solved using this device.

Such a brain was latched on to by the British government, and when Turing returned to Britain in 1938 he went to work at the government's Code and Cypher School. When war broke out, Turing moved with the School to new headquarters at Bletchley Park. Here, alongside his colleague Gordon Welchman, he used information from Polish cryptographers to design the 'Bombe', a computing machine that was used to crack the infamous German Enigma code.

For this and other groundbreaking work, Turing received an OBE in 1945. He could also have been credited with designing the first general-purpose electronic computer while working at the National Physical Laboratory in London, but the fact that there was a delay in the construction of this so-called 'Pilot Ace' meant that he missed out on that accolade.

Frustrated by the delay, Turing moved to Manchester University, where he began work on the idea of whether a machine can actually 'think' for itself, devising an experiment to see if a concealed machine could fool someone into thinking it was in fact another person. This so-called 'Turing test' is still used today to test how human-like a machine's intelligence is.

Sadly, Turing's incredible life achievements are marred by events following his admission to the police of a homosexual affair with a petty criminal. Tried and convicted for homosexuality, which was illegal at that time, Turing opted for hormone treatment as opposed to the alternative of a lengthy jail sentence. Tragically, in June 1954, he was found dead at his home in Wilmslow in Cheshire; beside him was an apple laced with cyanide. His mother maintains that it was a freak accident caused by chemicals used to plate cutlery. But the verdict was suicide.

The Modern Era

The accolade of the first general-purpose electronic computer went to the ENIAC (Electronic Numerical Integrator And Computer) in 1946. It was designed at the University of Pennsylvania by John Mauchly and J. Presper Eckert. The first version used punched-card programs and vacuum tubes and was a thousand times faster than previous electromechanical machines. This was a leap in computing power that no single machine has matched since.

Computers of the 1940s, such as ENIAC, filled a whole room, consuming as much power as many hundred modern PCs. The advent of silicon transistors in the 1950s (see page 382) radically changed the size of computers. Then, in 1971, Intel developed the first microprocessor – or central processing unit (CPU) – that does most of the computation in modern computers. This 4004 model contained 2,300 transistors, which was impressive at the time, but not a patch on the 2 billion it's possible to fit on today's chips.

Despite a reduction in size, computers in the early 1970s were still restricted to offices and factories, but two young computer geeks

THE LOEBNER PRIZE

This annual competition sees computer programs, so-called 'chatterbots', being judged on how human-like their conversation is, which is the essence of the test that Alan Turing first proposed in 1950.

Since the contest first started in 1990, no chatterbot has ever fooled the judges into thinking they're talking to an actual person. So the $100,000 prize remains at large. This will remain the case until the day when the judges can't distinguish between a program and a human, and the competition will end.

were about to change that. Steve Jobs and Steve Wozniak met while working for Hewlett-Packard in their summer holidays. Together they joined the Homebrew Computer Club in 1974, building computer systems alongside other keen enthusiasts. Realising there would be a market for computers in the home, they started designing a computer which they called the 'Apple 1' in Jobs's bedroom, and then built the wooden-framed machine in the garage of his family home. By 1976 they'd set up the Apple Computer Company, planning to sell their first computers at their computer club, but a local electronics store spotted the machine's massive commercial potential and put in a $50,000 order.

The Apple 1 took off, partly because it could be turned on at the press of just one button, and partly because it could be programmed in BASIC, which followed more simple rules than many other computer languages. Owners were also able to write their own programs, and a tape recorder could be plugged into the computer to store the programs. The first models sold for $666 – about $2,000 in today's money. Only 200 of the Apple 1 were built, as they were made by hand. Less than a hundred exist and they now sell for tens of thousands of pounds at auction.

In 2010, over 1 million of Apple's iPads were sold in the first month that the device went on sale. Tablets, laptops and netbooks are becoming so integral to our busy commuter lives that one wonders whether the desktop computer has seen its best days. Or it may be that the vast computing power of today's smartphones means tablets will one day become redundant. Soon, docking stations on the back of monitors and TV screens that transform your phone into a web-browsing computer. All your information will be stored securely on a remote server in 'the cloud', and all the computing brains you'll need will be on the smartphone in your pocket.

PACKET-SWITCHED NETWORKS

The key to the invention of the internet

Nominated by **Bill Thompson**, technology author and contributor to BBC news and *Digital Planet*

Since the invention of the telegraph, it had been possible to have multiple connections over the same link. But there was a problem: each channel on the link could only cope with one call at a time, hogging it until the call was disconnected. The invention of 'packet-switching' was revolutionary. It split traffic data into chunks which could be routed via several pathways and then reassembled at the destination.

Like many great ideas in history, packet-switching emerged independently in different research

> *"The best-known packet-switched network today is, of course, the internet. Without the underlying transport model of packet-switching, however, we would have been forced to rely on telephone-style switching and things would be very different online."*
> Bill Thompson

teams. In the early 1960s, two men were looking for a way to make data transmission more efficient and reliable. Donald Davies, from the UK National Physical Laboratory, and Paul Baran, from the US armed forces's think-tank RAND, independently realised that messages could be broken up into separate routed packets, which were assembled on receipt. This meant there was no need to establish a persistent connection between the sender and recipient, so data transfer could be much more efficient.

This was good news for the US armed forces, as it meant communications could continue to run, even in extreme circumstances, such as a nuclear attack. The first packet-switched network, known as 'ARPA-net', was developed by the military and used from 1969. While ARPA-net was incredibly useful for the military, it was the springboard for a global public network that would revolutionise the world – the internet.

THE INTERNET

Linking humanity

Nominated by **Dallas Campbell**, presenter on
BBC science show *Bang Goes the Theory*

*"In human history it's difficult to imagine another technology
which, in such a ludicrously short space of time, has so
fundamentally altered the way we behave as a species. Not
just how we communicate, but how we trade with each other,
how we conduct our politics, the way we fall in love, the way
we defraud and commit crime. It's a game-changer in every
sense: instant access to the world's knowledge, competing with
instant access to the world's nonsense on a level playing field,
reflecting and amplifying back at us what it is to be human.
Changes which once might have taken generations have
happened in a couple of decades. And this is just the start..."*
Dallas Campbell

In the 1960s, with the threat of nuclear war ever present, the US
government commissioned the armed forces's think-tank RAND to

HOW THE INTERNET WORKS

A group of computers form a computer network, linking up to other networks via numerous routes – just like groups of houses connect up to different towns via a number of different roads.

Your computer communicates with others via packets of digital data that are sent along wires or wirelessly by radio waves to a router or modem. This then sends the data via a telephone line or TV cable to a small local network, known as an internet service provider (ISP). Fibre-optic cables on land or on the seabed transfer the data internationally between ISPs.

research more efficient and reliable data transmission methods. Paul Baran came up with the idea of of splitting data into chunks that could be routed via several networks.

Baran wasn't the only one in the sixties to invent this 'packet-switching'. Donald Davies at the UK Physical Laboratory also came up with the same idea. But Baran was able to apply this concept to create, in 1969, the first packet-switched network ARPAnet, which then gave rise to the global public network, the internet.

The internet has revolutionised our lives, allowing access to information through search engines like Google, and enabling global communication through the likes of Facebook, Twitter and Skype.

The internet also has its downsides. The obvious ones are spam and hacking, but there is also the potential for more serious, long-term consequences. As Campbell points out: "For all its obvious benefits, will an increasingly connected world completely homogenise human culture? And how will it affect the way we behave towards each other and our planet?"

Only time will tell.

DID YOU KNOW?

Every month, 1.6 billion people on average surf the internet.

THE WORLD WIDE WEB

The advent of information sharing

Nominated by **Professor Jonathan Zittrain**, professor of internet law at Harvard Law School and Harvard Kennedy School, and professor of computer science at Harvard School of Engineering and Applied Sciences, whose books include *The Future of the Internet: and How to Stop It*

The brains behind the world wide web, Tim Berners-Lee, started out like many a software geek, building his first computer with not much more than an old TV, a M6800 processor and a soldering iron. At the time he was studying physics at Queen's College, Oxford. After graduating with a first-class degree and stints at a couple of telecommunication manufacturing companies, he went solo – working as an independent software engineering consultant for CERN (famous today for its atom-smashing Large Hadron Collider). It was while Berners-Lee was at CERN that he had his first ground-breaking brainwave.

Berners-Lee created a hypertext database system, called 'ENQUIRE'. In basic terms, hypertext is text on a computer that has links (hyperlinks) through to other documents. Berners-Lee developed the program to help researchers share and update information among each other. But little did he know how useful it would become a decade later when he developed one of the world's most revolutionary ideas of all time.

In March 1989, using the ENQUIRE database, Berners-Lee came up with the idea of a global hypertext system – the world wide web. He apparently thought of the name while sitting in the cafe at CERN, but it was to be another year before he and colleague Robert Cailiau presented the concept to manager Mike Sendall, who reportedly wrote "vague but exciting" on the proposal and let them go ahead with the project.

Berners-Lee went on to build the first web browser and web server. The first website was built at CERN and put online on 6 August 1991. Without Tim Berner-Lee's vision, everything

"The web taught us that being online didn't have to mean subscribing to a company's gated community of negotiated content. The beauty of the web was that it retained the basic atoms of the internet – universally addressable nodes of a network so that information and services could repose anywhere – and then built some extraordinary molecules. For the first time, information could be assembled together on a single page from multiple sources without any advanced agreements or planning among them – a link would suffice. The web is simply a protocol: a way to build pages so that others can see them. That alone brought down the centrally orchestrated content gardens of the day – CompuServe, Prodigy, Minitel, and so on. What an amazing network, where the insight of one man, in this case Tim Berners-Lee, could so rapidly transform the online world."

Jonathan Zittrain

DID YOU KNOW?

The average person who uses the internet visits more than 1,000 web pages every month.

from information storage to communication might be completely different today. But as Zittrain points out, a new vision could be just around the corner: "Tim Berners-Lee had amazing vision. And today's web invites successors."

SOCIAL NETWORKING

From the Well to Twitter, via Friendster and Facebook

Nominated by **Rory Cellan-Jones**, BBC technology correspondent (with more than 27,000 Twitter followers)

Around half of the UK now has a Facebook account. The number of active users globally is about 600 million, and that's over 8 per cent of the world's population. To put it in perspective, if all those people lived in one country, it would be the third most populated in the world.

With such global dominance, it's not surprising that Facebook creator Mark Zuckerberg now has an estimated wealth of $13.5 billion, which he's acquired in just a few short years.

"The web made a space where we could all be creators – and passive consumers. But it wasn't until social networking, in the form of Facebook and Twitter, that Tim Berners-Lee's original vision was realised: a place to share every detail of our daily lives, often mundane, but sometimes revolutionary, as we've seen in the popular uprisings in the Middle East."

Rory Cellan-Jones

Zuckerberg was always a bit of a genius, writing computer programs while still at school. In his second year at Harvard University in 2003, he wrote a program called 'CourseMatch', where users could pick classes according to other students' choices. He also created Facemash, a site where students could vote on the attractiveness of fellow students.

Facemash was intended as a bit of fun, but after some students complained, Zuckerberg was forced to make a public apology. At the time, other students were asking for the university to create an internal website of contact details and photos and Zuckerberg saw its potential. He launched the site from his university dorm room on 4 February 2004, calling it 'Facebook' – possibly inspired by his prep school student directory, which students nicknamed 'The Facebook'.

Long before Facebook, a successful virtual community existed, though. The Whole Earth 'Lectronic Link, aka the 'Well', was launched in 1985 and became best known for its internet forums. It was followed by many others in the 1990s, including Classmates. com, where users could track down former classmates with similar interests, create profiles and send messages. PlanetAll.com went a step further by recommending potential friends. But the big breakthrough was creating social networking sites that users could manage themselves. Friendster launched in 2002, followed by LinkedIn, MySpace and Bebo. In 2005, MySpace was getting more page views than Google.

That same year, three former employees of e-commerce business PayPal, created the video-sharing site YouTube. Today, every minute, 24 hours of video is uploaded to YouTube, while the 200 million users of microblogging site Twitter generate 65 million 'tweets' a day.

TWENTY MINUTES IN THE LIFE OF FACEBOOK

- Tagged photos: 1.3 million

- Event invites sent out: 1.5 million

- Wall posts: 1.6 million

- Status updates: 1.8 million

- Friend requests accepted: 1.97 million

- Photos uploaded: 2.7 million

- Comments: 10.2 million

- Messages: 4.6 million

As with so many social networks, which start on a small scale, aiming to service a particular community or institution, Twitter was created in 2006, initially as a service for employees of Odeo, a directory and search destination website for people to share podcasts.

With our lives so hooked into the likes of Facebook, LinkedIn, YouTube and Twitter, one wonders what institution or community is dreaming up the next big thing in social networking. As Cellan-Jones puts it: "It's been a very speedy and unpredictable revolution, with networks like MySpace rising and fading within a couple of years. For now, Facebook looks like becoming one of the most powerful forces in global lives, but there's no guarantee that another idea won't come along to transform the way we communicate."

INDEX

Note: page numbers in **bold** refer to illustrations.

CREDITS

Thank you to Robert Matthews for his help with writing and letting me pick his genius brain when needed. Thanks also to Emma Bayley for her writing and proofing, Nikki Withers for her fast-fingered typing, Graham Southorn and Will Gater for letting me tap into their vast space knowledge. Thanks to Jack Searle, Jennifer Ross, Leah Nedahl, Kerry Rodgers, Julie McFarlane, Sarah Jordan, Sam Shead, Michael Moran, Ruth Norris, Jenny Gramnes and Eloise Kohler for their help with research – and the Focus team for their support.

Plus, a big thank you to the experts, without which this book would never have existed: David Adam, Dan Ariely, Jim Al-Khalili, Stephen Baxter, Colin Blackmore, Sue Blackmore, Michail Bletsas, Liz Bonnin, Tim Boon, Donal Bradley, Andrea Brand, Richard Branson, James Caan, Dallas Campbell, Dermot Caulfield, Rory Cellan-Jones, Marcus Chown, Stuart Clark, Nicola Clayton, Frank Close, Andrew Cohen, Francis Collins, Brian Cox, Lewis Dartnell, Kay Davies, Richard Dawkins, Daniel Dennett, David Deutsch, Athene Donald, Marcus du Sautoy, James Dyson, John Emsley, Ross Ethier, Patricia Fara, Penny Fidler, Henry Gee, John Gribbin, Adam Hart-Davis, Dan Heaf, Tom Heap, Mark Henderson, Lesley Gavin, Ben Goldacre, Timandra Harkness, Nigel Henbest, Bill Jones, Michio Kaku, Spencer Kelly, Nick Lane, Aidan Laverty, Chris Lintott, Iain Lobban, Robert Matthews, Bill McGuire, Patrick Moore, Michael Mosley, Dave Musgrove, Elon Musk,

Steve Myers, Rahiel Nasir, Kim Nasmyth, Paul Nurse, Paul Parsons, Fred Pearce, Francis Pryor, Nick Psychogios, Martin Rees, Alice Roberts, Pat Roche, Steven Rose, Adam Rutherford, Danielle Schreve, Tara Shears, Seth Shostak, Simon Singh, Caroline Smith, Hal Sosabowski, Graham Southorn, Jem Stansfield, Ian Stewart, Chris Stringer, Ulrike Tillmann, Bill Thompson, Charlie Turner, Jimmy Wales, Mark Walport, Richard Wiseman, Lewis Wolpert, Yan Wong, Tom Welton, Luis Villazon, Carl Zimmer, John Zarnecki, Jonathan Zittrain.